Lecture Notes
in Control and Information Sciences 226

Editor: M. Thoma

Springer-Verlag London Ltd.

G. Cooperman, G. Michler and H. Vinck (Eds)

Workshop on High Performance Computing and Gigabit Local Area Networks

 Springer

ISBN 978-3-540-76169-3

British Library Cataloguing in Publication Data
Workshop on high performance computing and gigabit local
   area networks   -   (Lecture notes in control and Information
   sciences ; 226)
   1.High performance computing   2.Local area networks
   (Computer networks)
   I.Cooperman, Gene   II.Michler, G. O. (Gerhard O.), 1938-
   III.Vinck, Han
   004.3'5
ISBN 978-3-540-76169-3   ISBN 978-3-540-40937-3 (eBook)
DOI 10.1007/978-3-540-40937-3

Library of Congress Cataloging-in-Publication Data
A catalog record for this book is available from the Library of Congress

Typesetting: Camera ready by editors and contributors

69/3830-543210 Printed on acid-free paper

Preface

A workshop on "High Performance Computing and Gigabit Local Area Networks" was held from 29 April until 3 May 1996 at the Institute of Experimental Mathematics of Essen University, Germany. It was organized by Professor Gene Cooperman (Northeastern University, Boston) and Professor Gerhard Michler (Institute for Experimental Mathematics, Essen). The 32 participants came from Australia, Germany, the Netherlands, Spain and the United States.

The goal of this conference was to identify the interaction among parallel algorithms for numerical analysis and linear algebra, network protocols and parallel hardware. The meeting was motivated by demand to solve ever larger problems. While parallel computers are the obvious solution, it is unlikely that one architecture will optimally support all algorithms, or that one algorithm will fit all architectures.

Lectures on existing methods for computation with large matrices over gigabit networks were given in order to focus discussion, by providing a benchmark against which to measure other possibilities. By bringing together scientists working in computer networks, network protocols and numerical and linear algebra, the meeting has given new insights that would not be otherwise attainable. Only by interdisciplinary research through cooperation of electrical engineers, mathematicians and computer scientists it will be possible to determine the most efficient combinations of parallel algorithms, protocols and network hardware. This new and fruitful research area to which the 15 refereed research articles of these proceedings contribute.

At the end of these proceedings one will find the addresses of the workshop participants, along with the titles of the 22 original, invited lectures.

The conference was supported in part by the Deutsche Forschungsgemeinschaft, the Volkswagen Foundation, the University of Essen and IBM Deutschland GmbH. The editors of these proceedings thank these institutions and corporations for their generous help.

Essen, 29 Jan., 1997
G. Cooperman
G. Michler
H. Vinck

Contents

Introduction: High Performance Computing and the Technology of Switch-based Computer Networks

Gene Cooperman, G. Michler and H. Vinck

College of Computer Science
Northeastern University
Boston, MA 02115, USA
and
Institute for Experimental Mathematics
University of Essen
Ellernstr. 29, D-45326 Essen

Many important research problems in mathematics, science and engineering can only be solved by means of very demanding computations on supercomputers. Since these machines are very expensive, they are rare. Hence it is difficult to get access to them. The recent introduction of new high-speed network technologies may change this situation. In the near future, the combined computational power of a cluster of workstations interconnected by a high-speed local area network is likely to be comparable to that of a current stand-alone supercomputer. Such cost-effective and scalable facilities promise to become the workhorse of future large-scale scientific and engineering applications. Access to such facilities is particularly important for universities, as a more cost-effective substitute for expensive, traditional supercomputers.

This promising scenario motivated the workshop on "High Performance Computing in Gigabit Local Area Networks" at the Institute for Experimental Mathematics at Essen University. In order to survey the present state of the computer network technology and the computational demands in mathematics, science and engineering, a collection of experts were invited to participate, ranging from network technology through computer organization, parallel and distributed computing, computer science, numerical analysis, multimedia and scientific applications. K. Abdali of the NSF gave a broad perspective on these issues through his talk, "High Performance Computing Research under NSF Support".

It is the interplay between computational mathematics, computer science and the technology of computer networks which yields a fruitful new area of interdisciplinary research. This theme was expressed by most of the participants, and it received particular expression during the panel discussion, with panelists G. Cooperman, D. Du, A. Hoisie and X. Sun. Much of the push for break-throughs in developments of new hardware and software for high performance distributed computing will come from outside applications. In the past, many of the largest computations have been represented by science and engineering, such as the grand challenge problems of the American high performance com-puting initiative (HPCC). This workshop emphasized many of the second tier applications which have not been able to attract substantial distributed comput-ing resources until recently. Nevertheless, these multimedia and mathematical applications are part of the infrastructure of science, and have the potential to become more pervasive than current grand challenge applications. As an ex-ample of these close relations, I. Niemegeers pointed out in his lecture that the technical problems with respect to signalling and control for multimedia commu-nication in broadband ISDN also arise in large discrete parallel and distributed mathematical computations over switch-based high speed networks.

Some of the more interesting observations to come out of the workshop concern specifically these relationships between network infrastructure and ap-plications. G. Quintana and X. Sun pointed out how numerical analysts had already been forced to take account of hardware implementations for the even simpler architecture of single workstations, where cache design and virtual mem-ory could greatly affect performance. The result was the BLAS package (Basic Linear Algebra Subroutines) for which implementations of standard algorithms are being re-written for each hardware architecture, in order to fit better with those hardware parameters. From this, BLAS implementations for distributed computing architectures are also following in a natural manner.

It was observed that currently the network technology of an IBM SP2 super-computer, in particular its high performance switch, is superior to any other dis-tributed network used for clusters of workstations, e.g.: Ethernet, FDDI, Fibre Channel, HIPPI or ATM (Asynchronous Transfer Mode). However, A. Hoisie pointed out that the expensive, high-speed switches of today may simply be part of a continuum, and future technology may bring such high-speed switches into the mainstream as VLSI continues to make that technology cheaper. As one travels down that continuum, one finds experimental results reported at the meeting that show that distributed network computing over high-speed, local ATM networks promises to achieve similar performance at reduced cost, pro-vided that the higher level protocols, device drivers, and network interfaces are improved (see also [4]). D. Du pointed out that the newer networks are already fast enough to carry the data in many high performance computations, and that some of the important current issues are latency and bottlenecks in the interface between the computer and the network.

At the workshop, experiments in such high performance computations over distributed networks were also reported. Some of these experiments include benchmarks of mathematical algorithms for parallel matrix multiplication and parallel differential equation solvers. Such experiments feed back not only into further research on the communication technology and protocols of high-speed local area networks, but also into better mathematical algorithms that are designed specifically for efficiency in distributed computing. Such algorithms must take advantage of the special properties of the hardware: protocols, device drivers, network interfaces and switches. In particular, it is important to diagnose how each of these subsystems affects the other, and how they can work better together.

Algorithmic parallel implementations were also well represented. Special emphasis was given to the parallel numerical linear algebra package ScaLA-PACK [1] dealing with large dense and sparse real matrices. It has found applications in many areas of applied mathematics, sciences and solid and fluid mechanics.

Advanced and demanding parallel algorithms also exist in several areas of discrete mathematics, especially in group representation theory and parallel linear algebra over finite fields. The serious development of efficient serial algorithms for matrix computations over finite fields began with the construction of some large sporadic simple groups in the 1970's. The Cambridge school was especially productive, culminating in R. Parker's meataxe routines. This package has been further refined by K. Lux and his colleagues at RWTH-Aachen. Important problems in group theory and modular representation theory demanded the parallelization of these algorithms for large matrix computations over finite fields. This was begun by members of the Institute for Experimental Mathematics of Essen University [2]. In his lecture, J. Rosenboom described an application using dense matrices of dimension 206184 over the finite field GF(2).

Indeed, just as LINPACK has often been used for benchmarking individual computers, current implementations of distributed parallel linear algebra over high dimension have the potential for stress testing distributed computing architectures. Such benchmarks would test the efficiency of large memory applications, along with the high-speed data transport between processors necessary for effective computation. In particular, the implementations of the parallel linear algebra algorithms over finite fields or polynomial rings [5] help to create experiments checking the technical communication properties of a broadband computer network with 155 Mbit/s bandwidth and higher. As was noted in the talk by G. Havas, applications in which the cost of the memory dominate have an interesting economic property. Parallel executions of such applications are often cheaper than sequential executions, since a parallel execution terminates sooner, and so one is "renting" the large memory for less time. This situation motivated the choice of a shared memory computer by G. Cooperman and G. Havas, in their work describing a parallel methodology that may yield the first enumeration of the 8, 835, 156 cosets for Lyons's simple group, in its smallest degree permutation representation.

In conclusion, the efficiency of computer communication networks heavily depends on high speed communication protocols and network interfaces. Applications in distributed computing will reach their full potential only if there is sufficient discussion between applications programmers, network designers, and a variety of disciplines in between. The recognition of this need is exemplified by the emergence of the first textbooks on this subject [3]. The lectures and discussions at the meeting have inspired several fruitful international research collaborations on high performance computing in local and wide area high speed computer networks. It emphasises the interdisciplinary cooperation in distributed high performance computing by electrical engineers, mathematicians and computer scientists.

References

[1] J. Choi, R. C. Whaley, I. Dhillon, J. Dongarra, S. Ostronchov, A. Petitet, K. Stanley, D. Walker, R. C. Whaley: ScaLapack: A scalable Linear-Algebra library for distributed memory concurrent computers - Design Issues and Performance Computer Science Department Technical Report CS-95-283, University of Tennessee, Knoxville (1995).

[2] P. Fleischmann, G.O. Michler, P. Roelse, J. Rosenboom, R. Staszewski, C. Wagner, M. Weller: Linear algebra over small finite fields on parallel machines, Vorlesungen Fachbereich Math. Univ. Essen, Heft **23** (1995).

[3] S. Hariri: High performance distributed systems: concepts and design, Prentice Hall, London, 1997.

[4] M. Lin, J. Hsieh, D. H. C. Du, J. P. Thomas, J. A. MacDonald: Distributed network computing over local ATM networks, IEEE J. On Selected Areas in Communications, Vol. 13, No. 4, May 1995.

[5] G. O. Michler, R. Staszewski: Diagonalizing characteristic matrices on parallel machines, erscheint in Linear Algebra and its Applications.

High Performance Computing and Communication Research at the National Science Foundation

S. Kamal Abdali
National Science Foundation
kabdali@nsf.gov

ABSTRACT: This paper describes the High Performance Computation and Communications (HPCC) of the National Science Foundation (NSF). The rationale and organization of the US level HPCC program are outlined to present the context. Then the NSF's HPCC-related activities are presented in more detail.

1 Background

The US High Performance Computing and Communication (HPCC) program was launched in 1991. It operated as a congressionally mandated initiative from October 1991 through September 1996, following the enactment of the High Performance Computing Act of 1991. From October 1996, it continues as a program under the leadership of the Computing, Information, and Communications (CIC) Subcommittee of the US National Science and Technology Council's Committee on Computing, Information, and Communications (CCIC). The program followed a series of national-level studies of scientific and technological trends in computing and networking[1, 2, 3, 4]. It was clear that the advances in information technology would affect society in profound, unprecedented ways. The program was thus established to stimulate, accelerate, and harness these advances for coping with societal and environmental challenges, meeting national security needs, and in increasing economic productivity and competitiveness. Formally, the goals of the HPCC program are to:

- Extend US technological leadership in high performance computing and computer communications

- Provide wide dissemination and application of the technologies to spread the pace of innovation and to improve the national economic competitiveness, national security, education, health care, and the global environment

- Provide key enabling technologies for the National Information Infrastructure (NII) and demonstrate select NII applications

2 Program Participants and Components

The HPCC program at present involves 12 Federal agencies, each with its specific responsibilities. The participating agencies are: Defense Advanced Research Projects Agency (DARPA), National Science Foundation (NSF), National Aeronautics and Space Administration (NASA), Department of Energy, National Institute of Health (NIH), National Security Agency (NSA), National Institute of Standards and Technology (NIST), Department of Veteran Affairs, Department of Education, National Oceanic and Atmospheric Administration (NOAA), Environmental Protection Agency (EPA), and Agency for Health Care Policy and Research (AHCPR). The activities sponsored by these agencies have broad participation by universities as well as the industry. The program activities of the participating organizations are coordinated by the National Coordination Office for High Performance Computing and Communication (NCO), which also serves as the liaison to the US Congress, state and local governments, foreign governments, universities, industry, and the public. The NCO disseminates information about HPCC program activities and accomplishments in the form of announcements, technical reports, and the annual reports that are popularly known as "blue books" [5, 6, 7, 8, 9]. It also maintains the website *http://www.hpcc.gov* to provide up-to-date, online documentation about the HPCC program, as well as links to the HPCC-related web pages of all participating organizations.

The program currently has five components: 1) High End Computing and Computation, 2) Large Scale Networking, 3) High Confidence Systems, 4) Human Centered Systems, and 5) Education, Training, and Human Resources. Together, these components are meant to foster, among other things, scientific research, technological development, industrial and commercial applications, growth in education and human resources, and enhanced public access to information. Specifically, the goals of these components are the following (see Blue Book 97 [9] for an official description):

1. **High End Computing and Computation:** To assure US leadership in computing through investment in leading-edge hardware and software innovations. Some representative research directions are: computing devices and storage technologies for high-end computing systems, advanced software systems, algorithms and software for modeling and simulation. This component also supports investigation of "ultra-scale computing" ideas such as quantum and DNA-based computing that are quite speculative at present, but may lead to feasible computing technologies in the future, and may radically change the nature of computing.

2. **Large Scale Networking:** To assure US leadership in high-performance network components and services. The supported research directions include: technologies that enable wireless, optical, mobile, and wireline communications; large-scale network engineering, management, and services; system software and program development environments for network-centric computing; and software technology for distributed applications, such as electronic commerce, digital libraries, and health care delivery.

3. **High Confidence Systems:** To develop technologies that provide users with high levels of security, protection of privacy and data, reliability, and restorability of information services. The supported research directions include: system reliability issues, such as network management under overload, component failure, and intrusion; technologies for security and privacy assurance, such as access control, authentication, encryption.

4. **Human Centered Systems:** To make computing and networking more accessible and useful in the workplace, school, and home. The technologies enabling this include: knowledge repositories; collaboratories that provide access to information repositories and that facilitate sharing knowledge and control of instruments at remote labs; systems that allow multi-modal human-system interactions; and virtual reality environments and their applications in science, industry, health care, and education.

5. **Education, Training, and Human Resources:** To support research that enables modern education and training technologies. The education and training is targeted to produce researchers in HPCC technologies and applications, and a skilled workforce able to cope with the demands of the information age. The supported research directions include information-based learning tools, technologies that support lifelong and distance learning for people in remote locations, and curriculum development.

The original HPCC program components, in force from 1991 to 1996, were: 1) High Performance Computing Systems, 2) National Research and Education Network, 3) Advanced Software Technology and Algorithms, 4) Information Infrastructure Technology, and 5) Applications, and Basic Research and Human Resources. The new program component structure reflects a refocusing of the HPCC activities in view of the experience and progress of the last five years.

3 HPCC at NSF

As stated above, NSF is one of the 12 agencies participating in the HPCC program. In the total HPCC budget request of $1143M in FY 96 for all agencies, NSF's share is $314M. This represents nearly 10% of NSF's annual requested budget for FY 96. NSF's HPCC-related work spans across all of the five HPCC program components. The objectives of NSF's HPCC effort include 1) fundamental research in high-end computing, 2) technological advances in hardware and software that are prerequisite for HPCC applications, 3) development of national HPCC facilities and services so as to make HPCC accessible to scientific and industrial researchers, educators, and the citizenry, 4) creating partnerships among universities, research labs, and industry to develop advanced computational infrastructure for HPCC, and 5) training of a scientific work force conversant in HPCC.

HPCC research penetrates to some extent nearly all the scientific and engineering disciplines at NSF, and most of the work undertaken by NSF's Directorate of Computer and Information Science and Engineering is related to HPCC. Representative ongoing research topics include: scalable parallel architectures; component technologies for HPCC; simulation, analysis, design and test tools needed for HPCC circuit and system design; parallel software systems and tools, such as compilers, debuggers, performance monitors, program development environments; heterogeneous computing environments; distributed operating systems, tools for building distributed applications; network management, authentication, security, and reliability; intelligent manufacturing; intelligent learning systems; problem solving environments; algorithms and software for computational science and engineering; integration of research and learning technologies.

4   Large HPCC Projects

The HPCC program has led to several innovations in NSF's mechanisms for supporting research and human resources development. The traditional manner of funding individual researchers or small research teams continues to be applied for HPCC work too. But to fulfill HPCC needs, NSF has initiated a number of totally new projects, such as 1) supercomputing centers and partnerships for advanced computational infrastructures, 2) science and technology centers, 3) various "challenges", and 4) the digital libraries initiative. These projects are much larger than the traditional ones in scope of research, number of participating investigators, research duration, and award size.

4.1   Supercomputing Centers (SCs) and Partnerships for Advanced Computational Infrastructure (PACIs)

NSF SCs were actually started in 1985, before the congressional formulation of the HPCC initiative. But they greatly contributed to the momentum behind HPCC, and, since its launch, have served to advance its cause.

Currently NSF supports four SCs: Cornell Theory Center, Cornell University; National Center for Supercomputing Applications, University of Illinois at Urbana-Champaign; Pittsburgh Supercomputer Center, University of Pittsburgh; and San Diego Supercomputer Center, University of California–San Diego. They are working cooperatively in a loose "metacenter" alliance, producing the impression of a single supercomputing environment to the users. Together the SCs provide a wide spectrum of high performance architectures, such as traditional (vector) supercomputers, massively parallel processors, and networks of high performance workstations. The SCs cater to the supercomputing needs of US computational scientists and engineers. But their customers range from neophytes to demanding experts. The rapid pace of parallel architecture advance does not allow vendors much time to devote to software. Thus the SCs obtain essentially bare machines, and build operating software to make the machines usable by the research community.

The SCs go much beyond offering a service facility. They are aggressively engaged in software development, user education, and research collaboration. They have been very effective in spreading the parallel computation culture. They have introduced many innovative, influential software tools. For example, the web browser Mosaic has greatly popularized web use, has had revolutionary impact on information dissemination and acquisition, and has given rise to an entirely new web-based software industry. Another significant product is Cave Automatic Virtual Environment (CAVE). CAVEs are multi-user room-size virtual reality environments with 3-D projections and sound surrounding the users. CAVEs are being used in highly interactive applications such as drug design and medical imaging. There is also research on linking multiple remote CAVEs into a collaboratory. The SCs have contributed significantly to scientific visualization, math libraries, parallelization tools, and communication software. They have also led many standardization efforts, such as the development of the Hierarchical Data Format (HDF) as a file format standard, or the "national file system."

In 1996, the SC program is being replaced by the Partnerships for Advanced Computational Infrastructure (PACI) program. This program represents an extension and refinement of the metacenter alliance of the SCs. The program aims to create a nationwide high performance computational facility with participation by universities, research labs, state and local governments, and the private sector. The facility will help maintain US world leadership in computational science and engineering by providing access nationwide to advanced computational resources, promoting early use of experimental and emerging HPCC technologies, creating HPCC software systems and tools, and training a high quality, HPCC-capable workforce.

4.2 Science and Technology Centers (STCs)

STCs are large research projects each of which involves typically 50+ principal investigators from 10+ academic institutions, and also has links to the industry. The participants work together on interdisciplinary research unified by a single theme, such as parallel computing or computer graphics. STCs provide an evironment for interaction among researchers in various disciplines and across institutional boundaries. They also provide the structure to identify important complex scientific problems beyond disciplinary and institutional limits and scales, and the critical mass and funding stability and duration needed for their successful solution.

STCs carry out fundamental research, facilitate research applications, promote technology transfer through industrial affiliations, disseminate knowledge via visitorships, conferences and workshops, educate and train people for scientific professions, and introduce minorities and underrepresented groups to science and technology through outreach activities.

As a result of competitions that took place in 1989 and 1991, 25 STCs were established by NSF. The following four of those STCs are supported by the HPCC program: The Center for Research in Parallel Computation (CRPC) at Rice University; The Center for Computer Graphics and Scientific Visualization at the University of Utah; The Center for Discrete Mathematics and Theoretical Computer Science (DIMACS) at Rutgers University; and The Center for Cognitive Science at the University of Pennsylvania.

These STCs have contributed numerous theoretical results, algorithms, mathematical and computer science techniques, libraries, software tools, languages, and environments. They have also made significant advances in various scientic and engineering application areas. Their output has been impressive in quality, quantity, and impact.

For example, DIMACS devoted a "Special Year" in FY 95 to molecular biology. The series of research seminars and workshops during that year has contributed much to the acceleration of interest in molecular computing, DNA sequencing, and protein structure studies. Also at DIMACS, a concerted effort directed to the traveling salesperson problem led to a breakthrough on this topic.

Similarly, the work at CRPC has been instrumental in the development of parallel languages, compilers, libraries, systems, and tools. CRPC has provided key leadership in industry-wide standardization of High Performance Fortran (HPF), and the prototype compiler Fortran D produced at CRPC is serving as a model for industrial HPF compiler development. Other well-known languages and systems which CRPC pioneered or significantly contributed to include ParaScope, PVM, MPI, HPC++, and ADIFOR. CRPC is also a force behind the National High Performance Software Exchange, various "infoservers," solvers for problems posed over the net, and several other innovative knowledge sharing schemes. In addition, there is notable research produced at CRPC on parallel algorithms for physical simulation and optimization.

4.3 "Challenge" Programs

The term "Grand Challenges" was introduced very early during the HPCC initiative. It is meant to characterize problems that are generally recognized as very difficult based on current standards of science and technology. They are computationally intensive, and are bound to "stress test" high performance computing hardware and software. They also require multidisciplinary approaches. Moreover, their solutions promise high payoffs in terms of scientific advances and crucial societal and industrial benefits. Later, the term "National Challenges" was also introduced. "Grand Challenges" is now used for computationally intensive, usually scientific, problems, while "National Challenges" is used for informationally intensive, usually engineering, problems. The distinction is often blurred.

A "Grand Challenge" problem list published in the Blue Book 93 [5] includes the following: magnetic recording technology, rational drug design, high speed civil transports, ocean modeling, ozone depletion, digital anatomy, and design of protein structures. The list is intended to be only representative, as many more such problems exist, and are actually under investigation by various HPCC research teams.

NSF has funded about 30 high-visibility group effort projects under the "Challenge" program. First, there were two rounds of "Grand Challenges" competition in FY 1990 and 1992 that mainly emphasized scientific problems. Then, there was a competition called "National Challenges" in FY 1993 that emphasized engineering problems such as manufacturing, civil infrastructure and health care delivery. Finally, there was a competition called "Multidisciplinary Challenges" in FY 1994 that encompassed physical sciences, engineering, computer science, and problem solving environments.

Unleashing large masses of scientific talent in collaborative attack on problems has indeed brought results. The challenge projects, specially those existing for over 3 years, have been demonstrably very productive. The projects have yielded, for example, highly accurate simulations in biomolecular dynamics, pollutant chemistry, materials science, radiation, and cosmology. Advanced computational techniques have been developed for modeling ocean and atmosphere phenomena, oil reservoirs, combustion, etc. For example, the weather forecasting research has led to the development of a hurricane modeling system that has become operational in 1995. As another example, computational advances in fluid dynamics have provided new understanding of the mechanisms controlling the solar heliosphere.

Expected challenge outputs also include technological artifacts, such as robots that assist surgery or advanced microscopes that incorporate pattern recognition. Information systems are under development for health care delivery, civil infrastructures, education, manufacturing, etc. The problem solving environment research is developing some systems targeted to broad problem classes and some that are quite generic.

4.4 Digital Libraries

The Digital Libraries Initiative is a joint venture of NSF, ARPA, and NASA. Its purpose is to advance the technologies needed to offer information essentially about anything, anywhere, to anyone. A digital library is intended to be a very large-scale storehouse of knowledge in multimedia form that is accessible over the net. The construction and operation of digital libraries requires developing technologies for acquiring information, organizing this information in distributed multimedia knowledge bases, extracting information based on requested criteria, and delivering it in the form appropriate for the user. Thus, the Digital Libraries Initiative promotes research on information collection, analysis, archiving, search, filtering, retrieval, semantic conversion, and communication.

The initiative is supporting 6 large consortia consisting of academic and industrial partners. Their main project themes and their lead institutions are: scalable, intelligent, distributed library focusing on the California environment (University of California–Berkeley); multimedia testbed of earth and space science data (University of Michigan); "Alexandria Project," maps and pictures (University of California–Santa Barbara); integrated virtual library of networked information resources and collections (Stanford University); internet browsing library, starting with contents of scientific and engineering periodicals (University of Illinois); information digital video library, starting with videos from public and educational TV channels (Carnegie Mellon University).

5 Evaluation and Impact

General directions as well as clear objectives were defined for the HPCC program from the very beginning. Thus, evaluation is built into the program. Some objectives naturally lead to quantifiable measures of progress, such as computation speeds in gigaflops, communication bandwidth in gigabits, network extent in number of connected nodes, etc. On the other hand, there are qualitative aspects of progress, such as scientific breakthroughs, innovative industrial practices, societal penetration of knowledge and technology, quality of work force trained, etc.

There is much experience with the SCs and STCs as most of these have existed for 5–10 years. They have also been subjected to much formal evaluation, and have been found successful. All these projects have external advisory committees which review their operation, progress and planned future directions periodically. NSF also monitors their performance via periodic progress reports and site visits. In addition to evaluating them on a per project basis, NSF has conducted rigorous evaluation of the SC and STC programs themselves using various means. Results of two recent assessments by blue ribbon panels, with recommendations for the future, are published in [12] for SCs and [13] for STCs. Other parts of the NSF HPCC program have also produced impressive results.

Collaboration is emerging as an important theme of HPCC. Most HPCC programs have emphasized 1) multi-disciplinary, multi-investigator, multi-institutional research teams, 2) partnerships between academia and industry, and 3) cooperative, interagency sponsorship of research. So much success is attributable to the multidisciplinary approach that this mode of research has de facto become a requirement in "challenge-scale" projects.

The HPCC program's effectiveness is being monitored by various kinds of studies and review panels. There is consensus that the program has been successful on most fronts. Not only the year by year milestones for quantifiable progress have been met, but the activities undertaken by the program have had noticeable impact and have led to several significant, unanticipated beneficial developments.

6 Critique and Conclusion

The initiative has been less successful in the area of software systems and tools. No unifying model of practical parallel computing has emerged that encompasses all the dominant current architectures. The variety of architectures has therefore caused fragmentation of the software development efforts. Lack of suitable software has retarded the spread of parallel computation. High performance computing has been very effective in some niche civilian markets. For example, large data bases, commercial transaction processing, and financial computing have been found well-suited for the new high performance machines. But in spite of the successful exploitation of high performance computing in such specialized applications, parallel computation has not yet become the paradigm for the masses. As a result, the parallel computing industry is far from being a commercial success.

In scientific computing and computer science, the exploitation of HPCC is very uneven. This difference is generally accounted for by the nature of the subject applications. Some problems lend themselves naturally to HPCC because of the regularity in their underlying data structures. But many problems with inherently irregular, heterogeneous data remain daunting to parallelization efforts.

In several numerical and optimization areas, there is considerable progress in adapting computations to high performance machines. But in fields such as symbolic and algebraic computation, automated deduction, and computational geometry, practical parallelization efforts are still in initial stages, though there is a lot of theoretical research. With the emergence of standards such as MPI that have been implemented on most architectures, it now seems possible to have "quick and dirty" parallel versions of important applications in the above mentioned areas. In theoretical computer science, the common PRAM model is not sufficiently realistic for analysis of parallel algorithms, and there is currently no agreement on a general model that is sufficiently realistic. Farther out in the future, there are "ultra-scale" paradigms, such as molecular computing, that could radically change the nature of computing. At present these are highly speculative ideas, as yet undeserving to be called technologies, and aggressive research programs are needed to define and exploit their potential.

References

[1] *Supercomputers: Directions in Technology and Applications*, National Academy Press, Washington, D.C., 1989.

[2] *Toward a National Research Network*, National Academy Press, Washington, D.C., 1988.

[3] *Keeping the U.S. Computer Industry Competitive: Defining the Agenda*, National Academy Press, Washington, D.C., 1990.

[4] *Scaling Up: A Research Agenda for Software Engineering*, National Academy Press, Washington, D.C., 1989.

[5] *Grand Challenges 1993: High Performance Computing and Communications* ("FY 1993 Blue Book"), Federal Coordinating Council for Science, Engineering, and Technology, c/o National Science Foundation, Washington, D.C., 1992.

[6] *High Performance Computing and Communications: Toward a National Information Infrastructure* ("FY 1994 Blue Book"), National Science and Technology Council, Washington, D.C., 1993.

[7] *High Performance Computing and Communications: Technology for a National Information Infrastructure* ("FY 1995 Blue Book"), National Science and Technology Council, Washington, D.C., 1994.

[8] *High Performance Computing and Communications: Foundation for America's Information Future* ("FY 1996 Blue Book"), National Science and Technology Council, Washington, D.C., 1995.

[9] *High Performance Computing and Communications: Advancing the Frontiers of Information Technology* ("FY 1997 Blue Book"), National Science and Technology Council, Washington, D.C., 1996.

[10] *Evolving the High Performance Computing and Communications Initiative to Support the Nation's Information Infrastructure* ("Brooks- Sutherland Report"), National Research Council, National Academy Press, Washington, D.C., 1995.

[11] *From Desktop to Teraflop: Exploiting the U.S. Lead in High Performance Computing* ("Branscomb Report"), Pub. NSB 93-205, National Science Foundation, Washington, D.C., August 1993.

[12] *Report of the Task Force on the Future off the NSF Supercomputing Centers* ("Hayes report"), Pub. NSF 96-46, National Science Foundation, Arlington, VA.

[13] *An Assessment of the National Science Foundation's Science and Technology Centers Program*, National Research Council, National Academy Press, Washington, D.C., 1996.

Practical parallel coset enumeration

Gene Cooperman[*]and George Havas[†]
gene@ccs.neu.edu and havas@cs.uq.edu.au
College of Computer Science
Northeastern University
Boston, MA 02115, USA
and
School of Information Technology
The University of Queensland
Queensland 4072, Australia

Abstract

Coset enumeration is a most important procedure for investigating finitely presented groups. We present a practical parallel procedure for coset enumeration on shared memory processors. The shared memory architecture is particularly interesting because such parallel computation is both faster and *cheaper*. The lower cost comes when the program requires large amounts of memory, and additional CPU's allow us to lower the time that the expensive memory is being used.

Rather than report on a suite of test cases, we take a single, typical case, and analyze the performance factors in-depth. The parallelization is achieved through a master-slave architecture. This results in an interesting phenomenon, whereby the CPU time is divided into a sequential and a parallel portion, and the parallel part demonstrates a speedup that is linear in the number of processors. We describe an early version for which only 40% of the program was parallelized, and we describe how this was modified to achieve 90% parallelization while using 15 slave processors and a master. In the latter case, a sequential time of 158 seconds was reduced to 29 seconds using 15 slaves.

[*]Supported in part by NSF Grant CCR-9509783.
[†]Supported in part by the Australian Research Council.

1 Introduction

Coset enumeration programs implement systematic procedures for enumerating the cosets of a subgroup H of finite index in a group G, given a set of defining relations for G and words generating H. Coset enumeration is the basis of fundamental procedures for investigating finitely presented groups. Computer implementations are based on methods initially described by Todd and Coxeter [17]. A number of computer methods for coset enumeration have been described, including those of Cannon, Dimino, Havas and Watson [3] and of Havas [9]. The latter introduces some coset definition strategies which give substantial reductions in total cosets defined for some difficult enumerations.

Cannon *et al* [3] include comprehensive references to earlier implementations, while later descriptions may be found in Neubüser [13], Leech [12] and Sims [16]. Neubüser and Leech both provide useful introductions to coset enumeration while Sims, in a substantial chapter of his book, gives a formal account of coset enumeration in terms of automata and proves interesting results on coset enumeration behaviour.

The computational complexity of coset enumeration is not well understood. Even for a given strategy there is no computable bound, in terms of length of input and a hypothetical index, to the number of cosets which need to be defined in the coset enumeration process to obtain a complete table. (The existence of such a bound would violate unsolvability results for problems involving finitely presented groups.) Further, Sims [16, pp. 224-226] has proved that there is no polynomial bound in terms of maximum number of cosets for the number of coset tables which can be derived by simple coset table operations like those used in coset enumeration programs. This more practical result indicates that the running time of coset enumeration procedures in terms of space available may be very high.

We build a parallel architecture on TOP-C [5], which provides a lower master-slave layer that is loosely based on the STAR/MPI philosophy [4] (see section 3.1). The master-slave structure is convenient due to its ease of programming and debugging. The debugging is particularly easy because all messages are sequentialized through the master, allowing one to trace messages in order to more intuitively observe the operation of a parallel program.

Historically, TOP-C was preceded by STAR/MPI. STAR/MPI provides similar facilities for interactive languages. Implementations of STAR/MPI exist for Common LISP and for GAP (a general purpose language for group theory algorithms). STAR/MPI is based on distributed workstations linked by MPI (a well-known Message Passing Interface). TOP-C provides a C library for the same capabilities. Further, TOP-C can be used with threads on a shared memory architecture, or with MPI on a distributed memory architecture, without change to the end-user code.

We describe a practical parallel procedure for coset enumeration on shared memory processors. It is built upon an implementation of the Felsch type procedure by Schönert, which is the basis of the coset enumerator in GAP [14]. The ensuing parallel enumerator is particularly well-suited to difficult enumerations. We make an in-depth case study of a single enumeration, using different parallelization techniques. The analysis of this single case is preferred to a large test suite, since if affords more opportunity for in-depth analysis of specific performance factors, as the algorithm is varied.

2 Major considerations

2.1 Cost effectiveness: More CPU's make it cheaper

Even with the availability of much larger memories, space is still the most important factor in difficult coset enumerations. The determining factor on whether many hard coset enumerations can be completed or not is the space available for the coset table. In certain cases hundreds of megabytes of memory are required. Large amounts of memory are a scarce resource, and it is desirable to reduce the time during which it is needed. One computing environment which provides us with the capability of reducing the time for substantial memory use is the shared memory multiprocessor (SMP).

This was observed by Wood and Hill [19], who studied the situation analytically. In the course of the current research, the authors independently re-discovered such an analytical approach. Our calculation of the costs is presented below, with parameters chosen particularly to model the parallel coset enumeration described in this paper.

We can derive a rough formula that accounts for this phenomenon in which more CPU's are cheaper. Let us assume that a parallelized program has a constant overhead and then achieves a linear speedup in the number of processors. As we shall see, this is the case for the parallel coset enumerator. Let c be the number of CPU's, m be the amount of memory, and h be the constant overhead time. The time for the given calculation is then $t = h + k_1/c$ for an appropriate constant k_1. The cost of the architecture is then $k_2 t c + k_3 t m$, for appropriate constants k_2 and k_3. The cost of the architecture is minimized when

$$c = \sqrt{k_3 k_1 m/(k_2 h)}.$$

This implies that for every quadrupling of the memory needed for a calculation (assuming that the constant overhead remains fixed), we should double the number of CPU's to achieve the optimum cost.

2.2 Shared memory architectures

A shared memory architecture was preferred over a distributed memory architecture even though a distributed memory architecture might be more highly scalable. The reason for this is that the hardest coset enumerations use hundreds of megabytes of memory in a coset table. Rather than duplicate the coset table on each processor, requiring enormous amounts of memory, it is much more economical to have all processors use a single copy of the coset table, stored in shared memory.

While the concept of a shared memory architecture is intuitive, perhaps the issues of memory contention are not sufficiently appreciated. Our testbed (see section 4.2) uses 8-way interleaved memory. Hence, we might expect memory contention in the case where we used 16 processors (one master and 15 slaves). This would exhibit itself in less than linear speedup for larger numbers of slaves. In practice, we observed no such memory contention, even with 16 processors.

This observation demonstrates a natural advantage of master-slave parallel algorithms in shared memory architectures. Each slave has a busy period (executing a task) and an idle period (waiting for the next task). Since the slaves are asynchronous, any temporary memory contention is usually caused by an accidental overlapping of busy periods for too many slaves. Such a condition will automatically be eliminated by a natural load balancing. The memory contention will cause some slaves to slow down until the slaves become out of sync again. In our experiments, as the number of slaves increased, the ratio of the average slave CPU time to the total elapsed time decreased. Hence, memory contention was apparently decreasing with the number of slaves.

2.3 Basics of coset enumeration

The primary components of coset enumeration procedures can be considered in terms of three distinct phases: definition/deduction making, relator tracing and coincidence processing. These phases are described in detail in various ways in [3, 9, 12, 13, 16], to which the reader is referred for further information. Comprehensive studies of aspects of coset enumeration performance are included in [3, 9, 16]. These show enormous differences in behaviour for various individual problems, depending on strategy details. In order to facilitate this study we restrict ourselves to just one coset enumeration strategy and one class of particularly difficult problems.

A preliminary and tentative description of parallel coset enumeration is given in [1] from a different perspective. Here we show how coset enumeration can be effectively parallelized on an SMP machine.

The coset enumeration procedure with simplest interactions between the three phases is the Felsch method, which is described in [3, 9, 12, 13, 16]. The fact that these interactions are straightforward makes it the most suitable candidate for parallelization.

We use Martin Schönert's C-language implementation of the Felsch method as our base code. This program is a predecessor of the GAP coset enumerator and provides a clean, modular foundation for our developments.

With respect to the coset table, we observe that definition/deduction making is a write only operation, relator tracing is a read-only operation and coincidence processing is read-write. Coincidence processing (which is an instance of the union-find algorithm) is itself very fast, but may cause very substantial alterations to the coset table, while the other phases make either no changes (relator tracing) or minor changes (definition/deduction making alters at most two coset table entries). Hence we choose to parallelize the two easy phases, but we perform the coincidence processing sequentially. (This is another reason for choosing the Felsch procedure as our base, since it causes relatively few coincidences when compared with other methods.) We allow parallel writing with reading, but we ensure that all writing is done by the master to avoid write conflicts.

One type of coset enumeration which can be particularly space and time-consuming arises in studying groups satisfying identical relations, see [10, 18] for example. In this kind of situation the enumerations can be difficult and often have many defining relations, since many instances of the identical relations are used.

Standard Felsch-style coset enumerations alternate between definition making and relator tracing. All definitions and deductions are traced at all relevant relators at appropriate times. The relator tracing may lead to both deduction making and coincidence processing. In difficult enumerations the bulk of the time is often spent in the relator tracing.

3 Parallelization

3.1 Master slave architecture

The basic concept of the master-slave architecture is that tasks are generated on the master, and assigned to a slave. The result is computed on the slave and returned to the master. Different tasks are executed asynchronously. A SPMD (Single Program Multiple Data) style of programming is encouraged, in which the execution of a task on a slave depends only on the global variables of that slave, and in which the user program maintains the same data in global variables on all processors.

The parallelization of Schönert's sequential code is accomplished using three layers of code. The lowest layer is a simple implementation of message passing using the SGI (see section 4.2) system calls for threads. This layer exports routines such as

```
set_num_procs()
par_call()
finalize()
send_command()
get_result()
get_last_tag()
get_last_source()
is_master()
myid()
```

This layer contains all CPU-dependent parts, and can be replaced by a corresponding file for other host architectures.

The middle layer is a new implementation of a master-slave architecture in C. It is similar to the corresponding layer in the implementations of STAR/MPI [4] in GAP and Common Lisp. It exports the routines and global variables

```
init_master_slave()
master_slave()
ms_barrier()
master_slave_stats()
master_slave_trace
trace_cmd_format
```

The third layer of code is a simple modification of Schönert's sequential coset enumerator. In particular, it required writing new functions, get_task(), do_task(), and get_task_result(). These functions are parameters to master_slave().

A full description of the master-slave functionality is given in [4]. For this paper, it suffices to understand that the call to master_slave() will provide a parallel analogue of the following sequential code:

```
{ void *task;
  while (task = get_task(), NOTASK != task)
    get_task_result(do_task(task), task);     }
```

However in the parallel version, each invocation of do_task() operates on a potentially different slave, and different iterations through the while loop operate as if the iterations were overlapped. The master executes get_task() and get_task_result() on an as-needed basis as slaves either return results, or demand new tasks to execute. For related systems based on distributed memory and remote procedure calls, rather than the tasks described here, see [7]. The work in [11] also provides an alternative parallel interface.

The entire code development, including gaining initial familiarity with the SGI architecture and writing code for all three layers, required about two weeks of part-time activity by one person. This was helped by using the previous code from the master-slave layer of STAR/MPI as a model.

The time does not include the additional, and larger time needed for the many experiments in parallelization using the debugged code. The lowest layer (message-passing for threads) consists of about 200 lines of code. The middle layer (master-slave) consists of about 450 lines of code. The highest layer required adding about 250 lines of new code in an existing sequential program with 1300 lines.

3.2 Parallelization of coset enumeration

The first step in our parallelization is to do relator tracing in parallel. For some difficult examples, relatively few relator traces lead to either deductions or coincidences, so we simply record any trace which is productive, and delay any consequent deduction making or coincidence processing till a separate sequential phase. This does not damage the effectiveness of the coset enumeration procedure. (In the sequential phase we have the overhead of having to repeat the relevant traces, but this is relatively insignificant.)

In our very first experiment, we tried to distribute relator tracing on a relator by relator basis. However there was insufficient work done on each slave to justify the master/slave overheads. Thus, we need an adequate amount of work to do on each slave.

To increase the work per task, we then allocated individual definitions and deductions to slaves for tracing at all relators. (This is why long presentations provide good cases for parallelization.) This keeps slaves busy while there is a long enough work queue often enough. In standard Felsch methods this situation arises after a coincidence cascade, when long deduction queues are built up. In our case study that comprises about half the time.

When the work queue was small, many slaves were waiting idle. To improve on this, we added a parameter, PAR_THRESHOLD, for which 20 was found to be a good value. When a new work queue is created, the program operates in sequential mode (to avoid communication overhead). Only if the work queue grows in length to more than PAR_THRESHOLD, does the program switch over to the parallel, master-slave program. This is advantageous when the work queue is of length only one or two, which was often the case.

In order to increase the percentage of time when there is a long work queue, we alter the standard Felsch method, which defines one coset when there is no other work to do. Instead we define a number of cosets, namely PAR_THRESHOLD, in order to generate a long enough work queue to initiate parallel processing. This also does not damage the effectiveness of the coset enumeration procedure, but it does lead to local perturbations in coset definition order and minor variations in maximum and total coset statistics. This increases the parallelizable component of our coset enumeration procedure from about 40% to as much as 90% for our case study example.

4 A case study

4.1 Problem description

Our example is provided by a group we call T which plays an important role in Vaughan-Lee's proof [18] that Engel-4 groups of exponent 5 are locally finite. The group T is $G/\zeta(G)$ in [18, Lemma 11]. T is generated by a, b, c and satisfies the relations:

$a^5 = b^5 = c^5 = 1$;

$(a^r b^s c^t)^5 = 1$ for $r, s, t = \pm 1, \pm 2$;

$[a, b, a] = [a, b, b] = 1$;

$[a, c, a] = [a, c, c] = 1$;

$[b, c, c, c] = [b, c, c, b] = [b, c, b, c] = [b, c, b, b] = 1$;

$[c, a, b, a] = [c, a, b, b] = [b, a, c, a] = [b, a, c, c] = 1$;

$[b, a, c, [b, a]] = [c, a, b, [c, a]] = 1$.

We choose this example because it is difficult enough to provide a good test and it has enough defining relations to make it well-suited for our purposes. For the record, the subgroup $\langle ab, bc \rangle$ has index 3125 and requires a maximum of 25691 and total of 26694 cosets in a standard sequential Felsch enumeration.

4.2 Parallel facilities and experimental methodology

The work was done on Boston University's SGI Power Challenger Array. That computer consists of 18 R8000 CPU's, each running at 90MHz. The CPU's share 2 gigabytes of 8 way interleaved memory.

It was not possible to run our tests at a time when no one else was on the computer. Hence, our numbers had a small variability due to external loads. Nevertheless, the numbers were reproducible to within 10% during times of light loading. We ran each case several times, so as to distinguish outlier cases (due to momentary heavy loads on the machine) from the general body of results.

There was also an issue of which numbers to report as relevant. We took statistics for the total elapsed time, the CPU time on the master, the CPU time on a slave (actually the total CPU time of all slaves divided by the number of slaves), and load factors indicating how often the master and slaves were kept busy.

In reporting overall statistics, we chose to consider the total elapsed time and the average CPU time of a slave as the two figures of importance. The slave CPU time is also a measure of the elapsed time that the slave was active, as described below. One would prefer to measure the elapsed time that the slave is actively working on tasks, but the measurement would introduce too much overhead. Instead, we assume that the slave CPU time spent on a task is almost 100% of the elapsed time during that task. So, the slave CPU time is identified as the time for the "parallel portion" of the program. The difference of the total elapsed time and average slave CPU time was then labelled the time for the "sequential portion" of the program.

Note that one cannot point to a time when all slaves proceed in parallel (i.e. are busy at the same time). The decomposition into sequential and parallel portions works because each slave is busy for approximately the same amount of time, although the busy periods are not synchronous across slaves.

A slave on a lightly loaded machine would use almost all of the CPU time while it had a task, and it would use almost none of the CPU time while it was waiting for the next task. Hence, the CPU time on the slave was also considered to be a good measure of the elapsed time for the busy periods of the slave, i.e. the sum of the elapsed times during each task execution.

Thus, we used the average slave CPU time as an indication of the total elapsed time that the typical slave was active. Intuitively, this would seem to be a good measure of the portion of the program that operates in parallel, and the statistics bore this out. The average slave CPU time was almost exactly inversely proportional to the number of slaves.

4.3 Degree of Parallelization

Given the time, P, for the parallel portion and the time, S, for the sequential portion of the computation, as described above, one wants a figure of merit for the degree of parallelization. For N the number of slaves, we use a degree of parallelization,

$$\frac{N \cdot P}{S + N \cdot P}.$$

The quantity $N \cdot P$ should be (and is, in our case) approximately constant reflecting the total amount of parallelizable work to be done. The remaining quantity, S, reflects a period of time when most slave processors are idle, waiting for the master to provide additional tasks. Hence, the degree of parallelization is reflected by the ratio above.

The formula can be justified in terms of Brent's scheduling principle [2], which was originally derived for the PRAM model of parallelism. The principle says that an upper bound for the elapsed time for a parallel computation with p processors is $w/p + t$, where w is the total amount of work to be done (the total elapsed time required by a single processor), and t can be considered either as the total number of synchronization operations required or the elapsed time in the limit as infinitely many processors are applied.

Since in our case, $N \cdot P$ is constant as N varies, we identify $N \cdot P$ with the work w, which must also be constant. If we also identify S, the sequential portion, with t, then the formula reduces to saying: An upper bound on the elapsed time using N processors is $w/p + t = P + S$, which is indeed the case for us.

4.4   Results

Table 1 shows the performance of the SMP coset enumerator with parallelization of the relator trace phase in the context of the standard Felsch
method of defining one new coset at a time. We report measured cpu and
elapsed time in seconds. The load factor on the master never rose above
7%, indicating that the master was not near saturation and that still more
slaves could have been added (if processors were available). For reference,
the original sequential coset enumerator did its job in 158 seconds.

Number of slaves	Elapsed Time	Sequential CPU time	CPU per slave	Degree of Parallelization(%)
1 slave	176	104	72	41
2 slaves	143	107	36	40
3 slaves	128	104	24	41
4 slaves	119	101	18	42
5 slaves	108	94	14	43
10 slaves	98	92	7	43
15 slaves	108	103	5	42

Table 1: Single definition performance

With this initial level of parallelization, the degree of parallelization is
about 40% across the full range of processors. This indicates that only 40% is
parallelizable, and so the total speed up is limited to that. The work done by
the slaves ($N \cdot P$) is again approximately linear. A total of about 10200 tasks
were handled in parallel. The average load factor on a slave progressively
diminished from 70% with one slave down to 50% with 15 slaves. This would
indicate that as more slaves were added, slaves became idle more frequently,
as the deduction queue became empty faster.

Table 2 shows the performance of the SMP coset enumerator with multiple
definition making. Now the degree of parallelization increases from about
70% of the total execution time to as much as 90%. The work done by the
slaves ($N \cdot P$) is again approximately linear. The progressive increase in
the degree of parallelization (or equivalently the decrease in the sequential
portion of the time) is still somewhat surprising. This time about 45000 tasks
were performed in parallel. The maximum and total coset statistics varied
slightly, but were always within 100 cosets of standard Felsch. The average
load factor on a slave varied from 60% to 70% with no obvious pattern. Note
that the time of the sequential portion progressively decreased for this case
as the number of slaves increased. This is because the slaves can operate
almost all of the time.

Number of slaves	Elapsed Time	Sequential CPU time	CPU per slave	Degree of Parallelization(%)
1 slave	219	71	148	68
2 slaves	129	54	75	74
3 slaves	99	49	50	75
4 slaves	78	40	38	79
5 slaves	62	32	30	82
10 slaves	38	23	15	87
15 slaves	29	19	10	89

Table 2: Multiple definition performance

5 Future Work: Lyons-Sims Group

Of special interest is a project to attempt coset enumeration in the Lyons-Sims finite simple group by using the parallel coset enumerator. Recently, a full permutation representation of degree 9,606,125 for this group was explicitly constructed [6]. Although there is a slightly smaller permutation representation, this representation was used because it derives from an action on a conjugacy class, which has certain technical advantages. This was then used by Gollan [8] to reconstruct an explicit presentation, thus confirming Sims's original existence proof [15] for the group. These presentations will be used with the parallel coset enumerator to enumerate both the minimal degree permutation representation on 8,835,156 cosets and the 9,606,125 coset representation, and thus provide still other methods for showing the existence of the group and verifying presentations for it.

6 Conclusions

We have shown that parallel coset enumeration on shared memory archi-tectures can be effectively implemented and applied, using a master-slave architecture. The entire parallelization of previously sequential code was accomplished using only 900 lines of C code. Further, 650 of those lines implement lower layers of a general master-slave architecture. There were only 250 lines of new code that were specific to coset enumeration.

We have also shown (section 2.1) that for large memory calculations, such an architecture is both faster and *cheaper* than a single CPU computer with the same memory. Further, a master-slave architecture yields a natural division of a parallel program into a "sequential" and a "parallel" portion, where the parallel portion requires time inversely proportional to the number of CPU's.

Future work includes consideration of alternative enumeration procedures and alternative computer architectures. The results to date make it possible to consider enumeration of 10 million cosets in the Lyons-Sims group. Such enumerations will use a very large amount of memory, and thus a shared memory architecture will be required as much for cost effectiveness as for speed.

7 Acknowledgements

We acknowledge the provision of time by Boston University's MARINER project on the SGI Power Challenger Array, where this work was done. We also gratefully acknowledge use of Martin Schönert's coset enumeration program.

References

[1] S. Akl, G. Labontè, M. Leeder and K. Qiu, "On doing Todd-Coxeter coset enumeration in parallel", *Discrete Applied Math.* 34: 27–35 (1991).

[2] R.P. Brent, "The Parallel Evaluation of General Arithmetic Expressions", *J. of ACM* 21: 201–208 (1974).

[3] John J. Cannon, Lucien A. Dimino, George Havas and Jane M. Watson, "Implementation and analysis of the Todd-Coxeter algorithm", *Math. Comput.* 27: 463–490 (1973).

[4] G. Cooperman, "STAR/MPI: Binding a Parallel Library to Interactive Symbolic Algebra Systems", *Proc. International Symposium on Symbolic and Algebraic Computation (ISSAC '95)*, ACM Press, 1995, pp. 126–132.

[5] G. Cooperman, "TOP-C: A Task-Oriented Parallel C Interface", 5th *International Symposium on High Performance Distributed Computing* (HPDC-5), IEEE Press, 1996, pp. 141–150.

[6] G. Cooperman, L. Finkelstein, B. York and M. Tselman, "Constructing Permutation Representations for Large Matrix Groups", *Proc. International Symposium on Symbolic and Algebraic Computation (ISSAC '94)*, ACM Press, 1994, pp. 134–138.

[7] A. Diaz, E. Kaltofen, K. Schmitz and T. Valente, "A System for Distributed Symbolic Computation", *ISSAC'91 (Proceedings of the 1991 International Symposium on Symbolic and Algebraic Computation)*, ACM Press, 1991, pp. 323–332.

[8] H.W. Gollan, *A new existence proof for Ly, the sporadic simple group of R. Lyons*, Preprint 30, Institut für Experimentelle Mathematik, Universität GH Essen, 1995.

[9] George Havas, "Coset enumeration strategies", *ISSAC'91 (Proceedings of the 1991 International Symposium on Symbolic and Algebraic Computation)*, ACM Press, 1991, pp. 191-199.

[10] George Havas and M.F. Newman, "Applications of computers to questions like those of Burnside", *Burnside Groups*, Lecture Notes in Math. 806, Springer, 1980, pp. 211–230.

[11] N. Kajler, "CAS/PI: a Portable and Extensible Interface for Computer Algebra Systems", *Proc. of Internat. Symp. on Symbolic and Algebraic Computation (ISSAC-92)*, ACM Press, 1992, pp. 376–386.

[12] John Leech, "Coset enumeration", *Computational Group Theory*, Academic Press, 1984, pp. 3–18.

[13] J. Neubüser, "An elementary introduction to coset-table methods in computational group theory", *Groups — St Andrews 1981*, London Math. Soc. Lecture Note Ser. 71, Cambridge University Press, 1984, pp. 1–45.

[14] M. Schönert *et al.*, GAP – *Groups, Algorithms and Programming*. Lehrstuhl D für Mathematik, RWTH, Aachen, 1996.

[15] C.C. Sims, "The Existence and Uniqueness of Lyons Group", *Finite Groups '72*, North-Holland, 1973, pp. 138–141.

[16] C.C. Sims, *Computation with finitely presented groups*, Cambridge University Press, 1994.

[17] J.A. Todd and H.S.M. Coxeter, "A practical method for enumerating cosets of a finite abstract group", *Proc. Edinburgh Math. Soc.* 5: 26–34 (1936).

[18] Michael Vaughan-Lee, "Engel-4 groups of exponent 5", *Proc. London Math. Soc.* (to appear).

[19] D.A. Wood and M.D. Hill, "Cost-effective parallel computing", *IEEE Computer* 28: 69–72 (1995).

High Performance Computing over Switch-Based High-Speed Networks

Jenwei Hsieh and David H.C. Du

Distributed Multimedia Research Center and
Department of Computer Science
University of Minnesota
Minneapolis, MN 55455, U.S.A.

Summary. Communication between processors has long been the bottleneck of distributed network computing. However, recent progress in switch-based high-speed networks including ATM, HIPPI and Fibre Channel may be changing this situation. To utilize the aggregate computational power of clusters of workstations, We have especially concentrated our effort on maximizing the achievable throughput and minimizing the delay for end-users. In order to achieve better performance, we have investigated the following issues: i) ways to improve the performance of I/O subsystems and ii) porting PVM over high-speed network APIs. The typical communications between processors are going through a stack of network protocols. (i) represents our effort to improve the lower layer communication performance. However, the application (or user) level of performance may not be improved too much even if the lower layer performance is improved. Therefore, to improve user level performance it is required to bypass most of the network protocols in the communication stack and, hence, the overheads generated by them by porting applications (like PVM) directly over lower layer network protocols.

For (i) we discuss the details of how the I/O subsystems related to the network operations and several possible approaches for improving the maximum achievable bandwidth and reducing end-to-end communication latency. For (ii) we show some performance results of PVM over high-speed local area networks. Although the available bandwidth of high-speed network is much higher than the traditional LANs, the application-level performance, however, still lags behind the capabilities of the physical medium.

1. Introduction

Network computing offers a great potential for increasing the amount of computing power and communication facility for large-scale distributed applications. The aggregate computing power of a cluster of workstations interconnected by a high-speed local area network (LAN) can be employed to solve a variety of scientific and engineering problems. Because of the volume production, commercial workstations have much better performance to price ratio than Massively Parallel Processing (MPP) machines. With switch-based networks such as HIgh Performance Parallel Interface (HIPPI), Asynchronous Transfer Mode (ATM), or Fibre Channel, a cluster of workstations provides cost-effective, high-bandwidth communications. The advantages of network computing have already attracted many companies to use it as an alternative form of high-performance computing. A recent report shows that several companies in aeronautics industry utilize clusters of workstations for computational fluid dynamics processing and propulsion applications during off hours and weekends [7].

The typical methodology for network computing over clusters of workstations is based on a software framework that executes on participating workstations. The software framework provides a parallel programming environment that allows programmers to implement distributed algorithms based on a message passing model. Distributed applications utilize the computational power and communication facility of the cluster by using special libraries provided by the software framework. Those libraries usually support process management, synchronization, and message passing based on standard network protocols. Examples of such software framework are PVM [2, 5, 11], P4 [3], Express [9], Linda [4], and MPI [12].

Given an environment consisting of a cluster of workstations interconnected by a LAN, it is well-known among programmers that message passing facilities lack the performance found in distributed memory computers such as the Connection Machine CM-5 or the nCUBE which provides specialized high speed switch(es) or interconnect hardware. This is especially true about current slow speed (e.g., Ethernet) LAN. However, fast message passing is possible via three approaches. First, the change of the underlying network to a high speed network greatly increases the message passing speed. Second, improving the efficiency of device drivers of network interfaces. Third, bypassing the high-level protocol stack and using a lower layer protocol reduces overhead, and thus increases message passing speed.

In this paper we first present several approaches to improve the performance of a network I/O subsystem. We examine a network subsystem for an emerging high-speed network called Fibre Channel (FC). The objectives of this study is: 1) to understand the network subsystem and how it relates to network operations, 2) to evaluate and analyze the performance of such a subsystem, and 3) to propose possible approaches to improve application level throughput. The Fibre Channel device driver was monitored to understand how the network subsystem performs operations (basically it converts user-level requests into interface commands and sends the commands to the network interface). By using hardware and software performance monitoring tools, we were able to evaluate and analyze the performance of the network subsystem. Timing analysis is used to find performance bottlenecks in the network subsystem.

To further reduce the communication latency and exploit high speed networks, we enhanced a popular message-passing library, PVM, on clusters of workstations. The PVM message passing library was originally implemented using the BSD socket programming interface. The transport protocols used are the Transmission Control Protocol (TCP) and the User Datagram Protocol (UDP). Figure 1.1 shows how PVM was implemented on the BSD socket programming interface (on the right side of Figure 1.1). The main idea of improving PVM's message passing is to reduce the overhead incurred by the high-level protocols. The overhead incurred by the high-level protocols and the delay caused by the interactions with the host operating system can be varied by using different Application Programming Interfaces (APIs) which are available on different protocol layers [10]. In order to provide as close to optimal performance as possible, part of PVM's communication subsystem is re-implemented directly using two APIs, Hewlett Packard's Link Level Access (LLA) or FORE's ATM API, (on the left side of Figure 1.1) instead of the BSD socket interface. Since both APIs reside at a lower layer in the protocol stack, the overhead incurred when directly using these APIs is expected to be lower than the overhead incurred by using the BSD sockets programming API. We present these two implementation in Section 3. and 4. respectively.

The organization of the paper is summarized as follows. In Section 2, we present some experimental results on improving the network I/O subsystem for a Fibre Channel network. In Sections 3 and 4, we discuss our experiences of porting PVM

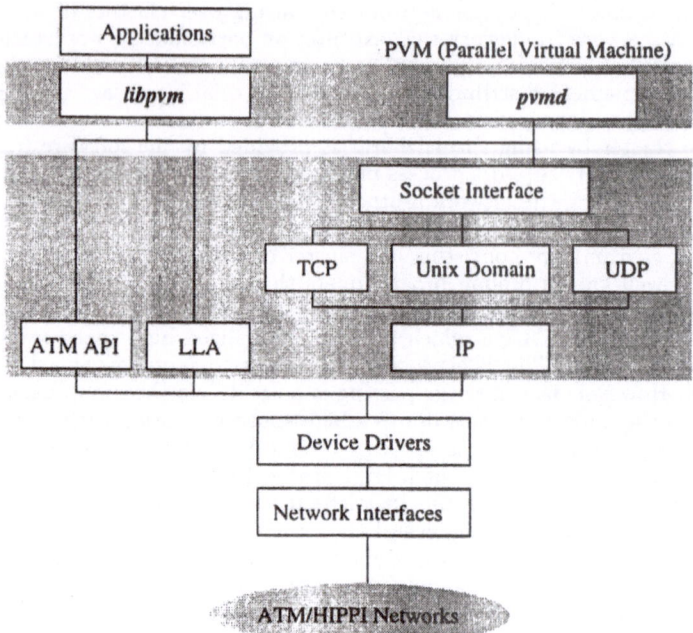

Figure 1.1. Protocol hierarchy of PVM's communication subsystem.

over ATM and HIPPI networks respectively. We offer some conclusions and future work in Section 5.

2. Performance Improvement of Network I/O Subsystem

The device driver is the link between the network interface and other system components. By analyzing the device driver, we were able to observe the data and control of each network operation and thus study the interactions among the various network subsystem components. A careful examination of the timing sequence of the network operations was used to evaluate the communication performance. Based on a simple timing analysis of these three services, we were able to identify several performance bottlenecks of the network operations. We found that application performance could be improved by simply modifying the device driver code. Since the device driver is a software component, it is relatively easy to modify (compared with hardware modification). Thus, the modification of existing device driver code represents one way to exploit and improve the capacity of the network subsystem.

The communication latency of the three classes of FC service can be partitioned into multiple timing phases. Each timing phase represents the time spent on the interactions of several components of the network subsystem. For example, a Class 1 sending process can be partitioned into the following four phases: entering the API, data movement, transmission over the physical network, and exiting the API.

The phases of entering and exiting the API involve the application programming interface, file subsystem (operating system), interrupt handling mechanism, and acquisition/release of the network interface. The data movement phase reflects the interactions among the memory subsystem, cache, system bus, I/O adapter, I/O bus (VME bus), and the DMA controller on the network interface. The physical transmission phase represents the communication behavior of the network interface.

2.1 Timing Analysis of Class 1 Write Service

In our FC test environment, two SGI workstations were used. The source side was a 4D/320VGX and the sink side was a 4D/340S. Each workstation was equipped with an Ancor VME CIM 250 Fibre Channel interface board which was connected to an Ancor CXT 250 Fibre Channel switch. The Fibre Channel interface (CIM 250) communicates with the host system using the VME bus. The interface has two dedicated FIFO buffers: Cmd_FIFO and Response FIFO. Both buffers (and all other on board registers) are mapped into the host memory space at startup time. The driver can then access the on board registers with normal memory read/write operations. It is the job of the I/O adaptor (IO3 board) to convert the memory read/write operation to VME read/write operations. When applications invoke a write system call, four major components of the I/O subsystem (host operating system, device driver, I/O adapter, and network interface) are required to cooperate with each other to complete the write operation.

The complicated interactions of a Class 1 write makes the timing analysis difficult. A straightforward approach is to divide the write operation into multiple timing pieces. However, there are many possible ways to partition the write operation. We chose to divide the write operation into several timing phases which represent special tasks being performed during the write operation. For example, there are two major tasks which need to be performed in order to transmit data to destination. One is the DMA operation across the VME bus. Another is the data transmission across the FC communication link. The timing phase before the DMA operation involves the system call interface. The phase after transmission across the FC link is the time spent checking the status and exiting device driver. One advantage of our *Task-oriented* partition is that it is easy to derive simple equations to approximate the experimental results. Since these four phases are executed sequentially, the write latency is equal to the sum of the time spent in each phase.

In Figure 2.1 the communication characteristics, based on the four phases of Class 1 write service, are presented. Figure 2.1(a) shows the write latency for message sizes ranging from 4 bytes up to 16 Kbytes. In Figure 2.1(b), the achievable user-level bandwidths are shown along with transmission rates of the DMA and FC phases. The transmission rate of the DMA phase represents the bandwidth of moving data from host memory to the interface via the VME bus. The transmission rate of the FC phase is the network bandwidth which the Fibre Channel provides. All data in the figure are measured by our experiments of monitoring lower level performance.

After collecting timing data, we derived simple equations to describe each phase. For each phase, there is a need to identify the time incurred by the subcomponents. For example, we observed the time for moving data from main memory to the interface could be characterized by the following three factors: *Word Latency* (T_{w_4}), *Block Boundary Latency* (T_{w_256}), and *Page Boundary Latency* (T_{w_4096}). User messages are stored in memory pages, and each memory page has 4096 bytes. T_{w_4096} represents the time incurred between the transmission of two consecutive memory pages (4096 byte page boundary) except for transmitting the last memory page.

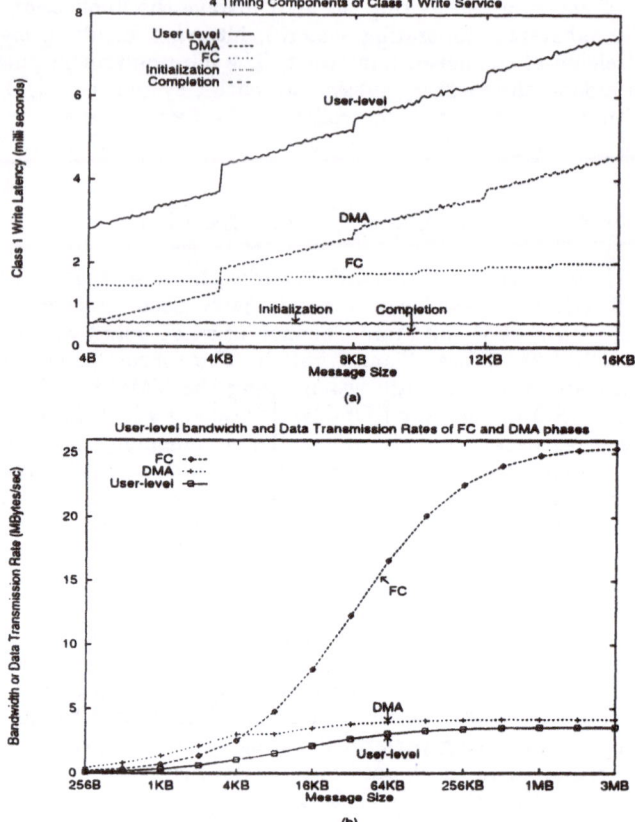

Figure 2.1. Timing analysis of Class 1 write (a) write latency (b) achievable bandwidth for different message sizes

The time for the last page of data is presented later. T_{w_256} is the time spent between the transmission of two consecutive memory blocks (VME block boundary.) T_{w_4} represents the overhead of moving one word from main memory to the network interface. The times observed using the VME analyzer for T_{w_4}, T_{w_256}, and T_{w_4096} are 0.2515 μsec, 1.76 μsec, and 636 μsec, respectively.

To further study the timing details of each component, we derived a simple timing equation for each phase. The equation contains one parameter, N which represents the user message size. Within the equations, all units of time are measured in microseconds (μsec.)

1. **Initialization phase** (T_{init_1w}) includes the time incurred by entering the Unix file system and initializing the interface.

$$T_{init\_1w} = 540 + N \times 0.000065$$

540 μsec represents the constant latency which does not depend on the size of the message. The second operand depends on the message size. This is the effect of network interface initialization.

2. **DMA phase** (T_{DMA_1w}) contains delays of physically locking the memory pages of the user buffer, preparing the address list, and moving data from main memory to the interface. There are two cases for different message sizes. For $N \leq 4$Kbytes,

$$T_{DMA\_1w} = 450 + T_{data\_1w} + T_{last\_1w},$$

where,

$$T_{data\_1w} = (\lceil \frac{N}{4096} \rceil - 1) \times T_{w\_4096} + (\lfloor \frac{N}{256} \rfloor - \lfloor \frac{N}{4096} \rfloor) \times T_{w\_256}$$

$$+ (\lfloor \frac{N}{4} \rfloor - \lfloor \frac{N}{256} \rfloor - 1) \times T_{w\_4}$$

$$T_{last\_1w} = 90 + ((N - 4) \bmod 4096) \times 0.1184$$

The T_{data_1w} term is the time incurred by moving data across the VME bus. We observed that the page boundary latency (except the last memory page) is 636 μsec. The total number of page boundaries equals $(\lceil \frac{N}{4096} \rceil - 1)$. Therefore, the total page boundary latency for transmitting N bytes of data is $(\lceil \frac{N}{4096} \rceil - 1) \times T_{w\_4096}$. When we applied a similar calculation to the block boundary and word latency, we got $(\lfloor \frac{N}{256} \rfloor - \lfloor \frac{N}{4096} \rfloor) \times T_{w\_256}$ and $(\lfloor \frac{N}{4} \rfloor - \lfloor \frac{N}{256} \rfloor - 1) \times T_{w\_4}$ respectively. The T_{last_1w} term is the time incurred between the last word of the DMA operation and the appearance of an acknowledgment interrupt. It depends on the size of the last user page. For $N > 4$ Kbytes,

$$T_{DMA\_1w} = T_{addr\_list} + T_{address\_1w} + T_{init\_DMA} + T_{data\_1w} + T_{last\_1w}$$

where,

$$T_{addr\_list} = 605 + \lceil \frac{N}{4096} \rceil \times 15,$$

$$T_{address\_1w} = (\lceil \frac{N}{4096} \rceil \times 3 - 1) \times T_{w\_4},$$

$$T_{init\_DMA} = 210 + N \times 0.008$$

The T_{addr_list} term is the time for physically locking user pages in memory. The time to physically lock a user page is around 15 μsec. There are a total of $\lceil \frac{N}{4096} \rceil$ memory pages for an N bytes message. The $T_{address\_1w}$ term is the DMA transfer time of the scatter-gather list. There are a total of $(\lceil \frac{N}{4096} \rceil \times 3 - 1)$ VME transfers. The interface preparation time for the first DMA operation is T_{init_DMA}, and it depends on the user message size.

3. **FC phase** (T_{FC_1w}) contains the time spent on setting up an FC connection, physically moving data across the network, and breaking down the connection.

$$T_{FC\_1w} = 1350 + \lceil \frac{N}{2048} \rceil \times 79.75$$

The per FC frame cost of moving data (2 Kbytes) from the sender to the receiver is around 79.75 μsec.

4. **Completion phase** (T_{comp_1w}) contains the time of unlocking physical memory pages, status checking, and returning from kernel space to user space.

$$T_{comp\_1w} = 295 + \lceil \frac{N}{4096} \rceil \times 5.9$$

295 μsec is a constant latency of the Completion phase. The cost to unlock a physical page is 5.9 μsec.

The calculated formula is very close to the experimental data. Mathematical modeling of the performance data allows us to identify possible communication bottlenecks and predict performance gains that should be possible after improvements are applied. For example, in the DMA phase, the page boundary latency is the major communication bottleneck. We will examine one approach to reduce the effect of the page boundary latency.

2.2 Approaches to Improve Communication Bandwidth

In this section, we discuss several possible approaches to improve the application level bandwidth. First, we describe special DMA mapping hardware and then compare the bandwidth improvements for different DMA segment sizes. The DMA segment size is the size of the data that can be moved in one DMA operation. Each DMA operation performed by the network interface uses multiple DMA block transfers.

In Section 2.1, the maximum application level bandwidth using the general purpose device driver is around 3.6 MBytes/sec. From Figure 2.1, we observed that the application level bandwidth is dominated by the DMA phase. The theoretical bandwidth of the VME bus is 40 MBytes/sec [13], however that bandwidth has never been accomplished. A reasonable bandwidth for a VME bus is around 25 MBytes/sec [13].

In Section 2.1, we observed the latency for writing one word (4 bytes) from main memory to the interface (T_{w_4}) is 0.2515 μsec, if the word is not at either page (4096) or block (256) boundary. The block boundary (T_{w_256}) and page boundary latencies (T_{w_4096}) are 1.76 and 636 μsec, respectively. If we use these parameters to calculate the achievable bandwidth, $4/T_{w\_4}$ is equal to 15.9 MBytes/sec. That is, without counting the latency caused by the block and page boundary, the maximum achievable bandwidth using DMA operation across the VME bus is 15.9 MBytes/sec. Considering the block boundary latency, we can get $256/(63 \times T_{w\_256} + T_{w\_256}) = 14.54$ MBytes/sec. When moving one page of data from main memory to the interface using DMA, the bandwidth is equal to $4096/(1007 \times T_{w\_256} + 15 \times T_{w\_256} + T_{w\_4096}) = 4.47$ MBytes/sec. That is, the maximum bandwidth which can be achieved in this environment is restricted to 4.47 MBytes/sec. Page boundaries are responsible for 70% of the total latency.

The page boundary latency is the largest bottleneck in the DMA operation. One approach to improving communication bandwidth is to increase the memory page size of the host system. However, the memory page size is fixed within the kernel and would be impossible to change. Another approach is to allocate the DMA buffer in physically contiguous memory, however this is not possible without copying the buffer in the kernel. The overhead of the copy makes this alternative undesirable.

Fortunately, SGI's IO3 adapter provides DMA mapping hardware which translates physical addresses to/from "bus virtual addresses" by using DMA mapping registers. Bus virtual addresses are a special range of addresses. When the operating

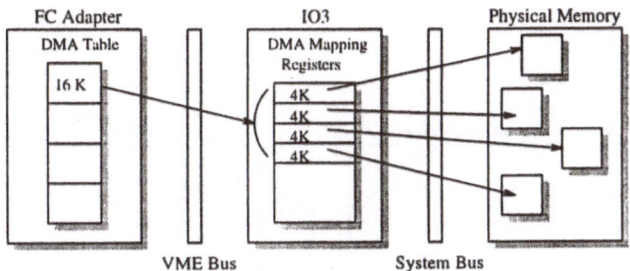

Figure 2.2. DMA mapping scheme of IO3 adapter

system transfers to/from one of these addresses, it knows that it is using registers on the IO3 board. Each DMA mapping register can map one page of memory (4 Kbytes). We can use this facility to prepare a larger DMA segment for the network interface. For example, to prepare a 16 Kbyte DMA segment, four DMA mapping registers are used to translate contiguous bus virtual addresses to four separate physical memory pages as shown in Figure 2.2. From the network interface point of view, these four physical pages are treated as a contiguous 16 Kbyte segment.

The FC device driver was modified to use this DMA mapping facility. After several simple experiments, we found that the segment boundary latency for 4 Kbyte and 8 Kbyte DMA segment sizes are 636 and 798 μsec, respectively. For DMA segment sizes greater than or equal to 16 Kbytes, the segment boundary latency is 1154 μsec. Therefore, the overhead of the DMA phase can be reduced by increasing the DMA segment size. For example, when moving a 16 Kbyte user message with 4 Kbyte DMA segments, there are four segment boundary. The latency equals $636 \times 4 = 2544\mu$sec. But with 16 Kbyte DMA segment size, the boundary overhead is 1154 μsec, it is an improvement of 1390μsec.

In the next experiment, we measured the application level bandwidth for six different DMA segment sizes ranging from 4 Kbytes to 1 Mbyte. The timing equations derived in Section 3.2 are also adjusted to accommodate the changes to the segment boundary latency and segment size. The timing equations were applied to predict the expected bandwidth which we then compared to the measured bandwidth. In Figure 2.3, we present the results. For the different DMA segment sizes, we calculated the "expected bandwidth" using the modified timing equations. The "original bandwidth" represents the performance of the network subsystem with the original device driver. The measured maximum bandwidth for six DMA segment sizes are 4.52, 5.14, 5.56, 5.81, and 6.06 MBytes/sec. For the 1 Mbyte DMA segment size, this resulted in a 75% application level bandwidth improvement.

2.3 Improving Communication Latency

The communication latency is critical for distributed applications such as distributed network computing and real-time multimedia applications. In this section, we first examined two possible approaches for reducing the Class 1 write communication latency. The basic idea for reducing the communication latency is to overlap some of the four phases of Class 1 Write. This will reduce the time that the interface and the driver spend interacting with each other. Then, we further discuss other possible approaches for reducing the communication latency.

2.3.1 Overlapping The FC and Completion Phases.

After the completion of the DMA phase, the device driver needs to unlock the physical memory pages. The device driver used in Section 2.1 unlocks the physical

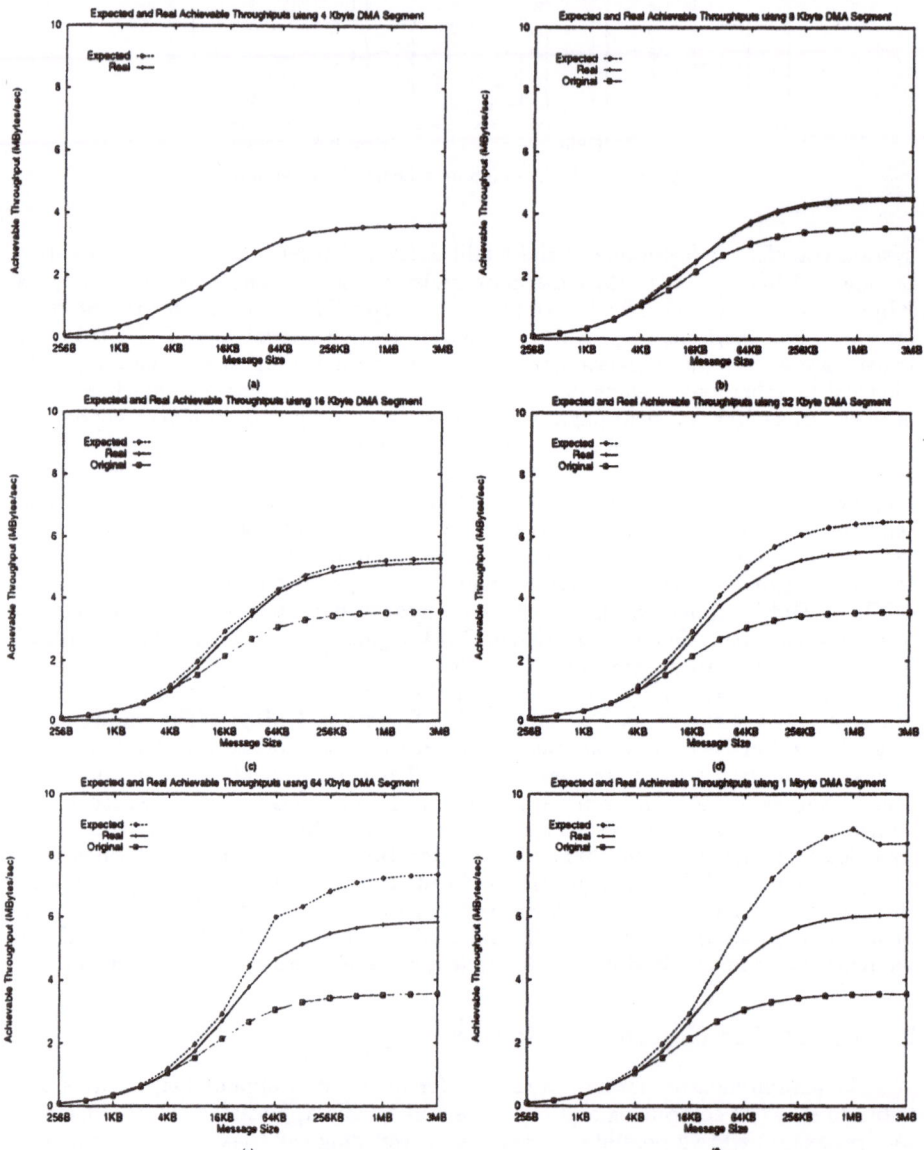

Figure 2.3. Comparison of expected and user achievable bandwidth when varying DMA segment size from (a) 4 Kbytes, (b) 8 Kbytes, (c) 16 Kbytes, (d) 32 Kbytes, (e) 64 Kbytes, and (f) 1 Mbytes

pages after the FC phase. However, these physical pages can be released right after the DMA phase instead of after the FC phase. The advantage of this approach is that it overlaps the FC phase and the Completion phase. The device driver can unlock the physical pages at the same time the network interface transmits the data across the Fibre Channel link. This approach reduces the overhead of releasing the memory pages. The amount by which the overhead is reduced depends on the number of memory pages which need to be unlocked. After implementing this concept in the device driver, the time spent on the Completion phase for a Class 1 write is a constant (190 μsec) instead of $(295 + \lceil \frac{message\ size}{4096} \rceil \times 5.9.)$

Table 2.1. Timing reduction for Class 1 write using overlap of FC and Completion phases

Message Size	4 bytes	64 bytes	1 Kbytes	64 Kbytes	1 Mbytes
Original	2837	2834	3218	21396	296039
Overlapping	2715	2727	3100	21147	294650
Reduction	122	107	118	249	1389
Percentage	4.3%	3.8%	3.7%	1.2%	0.5%

Table 2.1 shows the results of applying this overlapping approach. With a 4 byte Class 1 write, a reduction of 122 μsec was observed. For Class 1 reads, there is no difference after applying this approach because there is no FC phase in the read service.

2.3.2 Delayed Status Checking.

During the interactions between the device driver and the network interface, it is very likely that the network interface will receive interface commands in correct sequence and the device driver will get responses from the network interface in proper order. The responses from the network interface are the status of operation results or failures. Therefore, the second approach is to delay the status checking as late as possible.

For the four phases of Class 1 Write, the device driver sleeps and waits for the responses from the network interface three times. It waits for acknowledgment after each of the first three phases. Our approach is to delay the status checking for the first two responses until the device driver receives the third one. This approach overlaps the actions of the device driver and the interface by issuing more driver commands without waiting for the intermediate results from the interface. This *success-oriented* approach has another advantage. It eliminates four context switches. One disadvantage of this approach is that the delayed status checking may increase the user-level response time when there are errors detected by the interface.

Table 2.2. Timing improvement of Class 1 write using delayed status checking

Message Size	4 bytes	64 bytes	1 Kbytes	64 Kbytes	1 Mbytes
Original	2837	2834	3218	21396	296039
Overlapping	2715	2727	3100	21147	294650
Delayed	2385	2390	2777	20823	292607
Reduction	452	444	441	573	3432
Percentage	15.9%	15.7%	13.7%	2.7%	1.2%

Table 2.2 shows the results of implementing the delayed status checking approach. A 4 bytes Class 1 write had a timing reduction of 452 μsec. This overall

latency improvement comes from the reduction of the overhead associated with the Initialization and the DMA phases.

3. Enhanced PVM Communications on ATM Networks

In this study, we sought to achieve high performance (e.g., low latency, high bandwidth) not only by implementing PVM on a high speed medium, such as ATM, but also by minimizing possible sources of overhead. The overhead incurred by the high-level protocols and the delay caused by the interactions with the host operating system can be varied by using different Application Programming Interfaces (APIs) which are available on different communication layers. Several previous studies have found a significant portion of communication overhead occurring due to these interactions [6, 18, 19].

The PVM message passing library was implemented using the BSD socket programming interface. The transport protocols used are the Transmission Control Protocol (TCP) and the User Datagram Protocol (UDP) transport protocols. In order to provide as close to optimal performance as possible, in our study, PVM was implemented directly over the ATM Adaptation layer protocol via the Fore Systems' ATM API instead of the BSD socket interface.

In our study, we provided two main enhancements to the existing communications facilities. Since the Fore Systems' ATM API provides only "best-effort" delivery and no flow control, we implemented an end-to-end protocol which provides cell retransmissions as well as imposes flow control. We also took advantage of the inherent multicasting capability provided by the ATM switch to significantly improve upon existing PVM multicasting functionality. Note that multicasting may not be inherent to ATM. However, it is provided by the Fore Systems' ATM Switch.

3.1 Application Programming Interface

In Section 1., Figure 1.1 depicts the protocol stack from the application layer to the network transport (ATM) layer. In this study, we sought to minimize unnecessary overhead, and hence chose to implement PVM on the Fore Systems' ATM API rather than the BSD socket interface.

The Fore Systems' ATM API library routines support the client-server model. Consistent with ATM specifications, a connection (Switched Virtual Circuit or Permanent Virtual Circuit) must be established before data can be exchanged between a client and server. Typical connection-oriented client-server interactions may then proceed. The Fore Systems' user-level ATM library routines provide a socket-like interface. Applications can select the type of ATM AAL to be used for data exchange. In the Fore Systems' implementation, AAL Types 1 and 2 are not currently supported by Series-200 interfaces, and Type 3 and Type 4 are treated identically. Therefore, we may simply use AAL 4 instead of AAL 3/4 throughout this paper.

Bandwidth resources are reserved for each connection. Resource allocation is based upon the following three user specifications: (i) peak bandwidth, the maximum data injection rate which the source may transmit, (ii) mean bandwidth, the average bandwidth over the lifetime of the connection, and (iii) mean burst length, the average amount of data sent at the peak bandwidth. The network control function will compute the chances that the requested connection will create buffer overflow (cell loss) and consequentially accept/reject the connection request.

If the connection is accepted, an ATM VPI and VCI is allocated by the network. The device driver associates the VPI/VCI with an ASAP which is in turn associated with a file descriptor. Bandwidth resources are then reserved for the accepted connection. The network then makes a "best-effort" attempt to deliver the ATM cells to the destination. A "best-effort" attempt implies that during transmission, cells may be dropped depending on the available resources. End-to-end flow control between hosts and cell retransmissions are left to the higher layers.

3.2 PVM Communications: Existing and Enhanced

3.2.1 PVM Communication Facilities.

PVM interprocessor communication is based upon TCP and UDP. TCP is a stream-oriented protocol. It supports the reliable, sequenced and unduplicated flow of data without record boundaries. UDP is a connectionless datagram protocol which is conceptually similar to conventional packet switched networks. Messages delivered via UDP sockets are not guaranteed to be in-order, reliable or unduplicated.

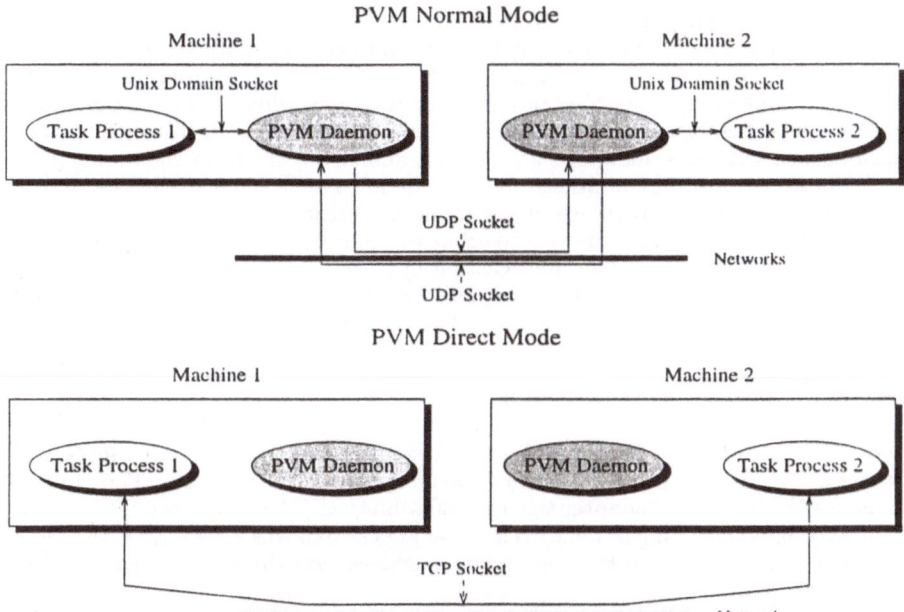

Figure 3.1. Comparison of PVM Normal and PVM Direct modes

PVM provides two types of communication modes, PVM *Normal* mode and PVM *Direct* mode. In the Normal mode, in order for a source task to communicate with a remote task, it must first communicate through a Unix domain socket to its local pvmd daemon. The local pvmd daemon of the source task then communicates through a UDP socket to the remote pvmd. The remote pvmd then communicates locally to the destination task through a Unix domain socket. Thus two TCP connections and two UDP connections are required for bi-directional communication between any two communicating application processes. In the Direct mode, PVM sets up a direct TCP connection between the two communicating processes or tasks.

The detailed transmission facilities of the Direct and Normal modes are hidden from the end-users. Figure 3.1 depicts these two modes.

The advantage of the Direct mode is that it provides a more efficient communication path than the Normal mode. A previous report [10] observed more than a twofold increase in communication performance between the two modes. The main reason PVM provides the Normal mode, despite its lower performance, is because of the limited number of file descriptors some Unix systems provide. Each open TCP connection consumes a file descriptor. Some operating systems limit the number of open files to as few as 32. If a virtual machine consists of N hosts, each machine must have $N - 1$ connections to the other hosts. Thus the drawback of the Direct mode is its limited scalability [11].

In order to enhance the performance and functionality of the existing PVM communication facilities, we made several changes to the existing PVM platform. First, we bypassed the BSD socket interface and directly implemented PVM on the lower level Fore Systems' ATM API. The second change we made was to improve PVM's multicasting capabilities. PVM assumes the underlying network cannot support multicast. We capitalized on the inherent multicasting capabilities of ATM, and re-implemented a more efficient multicasting operation.

3.2.2 Multicasting Protocol.

Efficient support for multicasting is important because multicast data flow patterns are often found in parallel programming applications. It is also important in a virtual machine environment because the host machines which comprise the virtual machine must regularly be coordinated, i.e., updated periodically when the virtual machine configuration changes, e.g., a host is added/deleted from the host pool.

PVM implements the multicast function by invoking a sequential series of send primitives. By taking advantage of the inherent multicast nature of ATM, we re-implemented the multicast function to occur as a parallel send to multiple receivers. Our re-implementation of the multicasting operation is the same as the original PVM except the N serial sends are replaced by a simultaneous send to N remote receiving hosts.

3.3 Performance Measurements

The ATM environment consists of Fore Systems's host adapters and local area switches. The host adapter was a Series-200 interface for the Sun SBus. The physical media for the Series-200 adapter was the 100 Mbits/sec TAXI interface (FDDI fiber plant and signal encoding scheme). The local area switch was a Fore ASX-100. Four Sun Sparc 2 machines and two Sun 4/690 machines were directly connected to the Fore switch.

An echo program is used for measuring the end-to-end communication latency between two machines. In this program, a client sends a M-byte message to a server and waits to receive the M byte message back. The client/server repeats this interaction N times. The round trip timing for each iteration in the client process is collected. The timing starts when the client sends the M byte message to the server, and ends when the client receives M bytes of the response message. Thus the problem of synchronizing clocks in two different machines is avoided. The communication latency for sending a M-byte message can be estimated as half of the total round-trip time. The communication throughput is calculated by dividing $2 \times M$ by the round-trip time (since $2 \times M$ bytes of message have been physically transmitted).

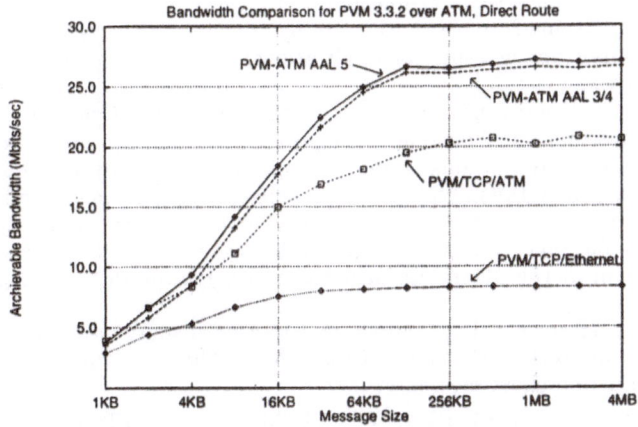

Figure 3.2. Direct mode: Bandwidth as a function of message size

Table 3.1. Direct Route

PVM Environment	r_{max} Mbits/sec	$n_{1/2}$ Bytes	t_0 μsec
PVM-ATM using AAL5	27.202	7867	1905.5
PVM-ATM using AAL4	26.627	8239	1903
PVM/TCP/ATM	20.826	7649	1839
PVM/TCP/Ethernet	8.312	1945	1541

3.3.1 End-to-End Performance.

We measured the performance for four different PVM platforms. PVM-ATM (AAL4 and AAL5) refers to the implementation of PVM directly on the Fore Systems' ATM API with the appropriate adaptation layer. PVM/TCP/ATM refers to the implementation of PVM on the BSD socket programming interface on an ATM network. PVM/TCP/Ethernet refers to the implementation of PVM on the BSD socket programming interface on an Ethernet network.

Figure 3.2 shows bandwidth as a function of message size for messages using the Direct route. PVM-ATM (AAL5) achieves the highest maximum bandwidth of 27.202 Mbits/sec. PVM-ATM (AAL4) achieves close to PVM-ATM (AAL5) bandwidth measurements, i.e., their maximum bandwidth measurements are within 0.6 Mbits/sec of each other. PVM/TCP/ATM achieves a maximum bandwidth value of 20.826 Mbits/sec. And PVM/TCP/Ethernet achieves a maximum bandwidth measurement of 8.312 Mbits/sec. From these results, we conclude that when PVM bypasses TCP and directly uses the ATM API a rather significant performance gain occurs of approximately 6 to 7 Mbits/sec. Again we observe a significant performance gain when using ATM as opposed to Ethernet.

The maximal achievable throughput is bounded by the speed of the TAXI interface, 100 Mbits/sec. In our previous study [10], we observed the maximum achievable throughput to be 46.08 Mbits/sec. In this study, we observed the maximum achievable throughput of PVM-ATM (AAL5) to be 27.202 Mbits/sec. Thus the overhead occurs at two levels: the end system and ATM interface (software and hardware) limits the throughput to 46.08 Mbits/sec, and the overhead from PVM limits the maximal throughput of PVM-ATM to 27.202 MBits/sec.

Table 3.1 shows the above described measurements in terms of the following three performance metrics.

1. r_{max} (*maximum achievable throughput*) : the maximum achievable throughput which is obtained from experiments by transmitting very large messages. This is an important measure for applications requiring large volumes of data transmissions.

2. $n_{1/2}$ (*half performance message length*) : the message size needed to achieve half of the maximum achievable throughput. This number may not be compared with the corresponding numbers from different hardware and software configurations, since the maximum achievable throughput may be different for different configurations. This measure provides a reference point that shows the effect of message sizes on the achievable throughput. Users can observe more than half of the maximum achievable throughput with messages larger than $n_{1/2}$.

3. t_0 (*round-trip latency*) : the time required to send a short message of 16 bytes to a receiver and receive the same message back. This is an important measure when transmitting short messages.

PVM-ATM (AAL5,AAL4), PVM/TCP/ATM, and PVM/TCP/Ethernet provided decreasing values for t_0, respectively. The greatest time difference occurs between using ATM or Ethernet. The overhead, in terms of latency, for the ATM network is thought to be primarily caused by the device driver. It is believed that the firmware code for Ethernet has been fine-tuned for better communication latency [10].

3.3.2 Multicasting Measurements.

Over ATM, we measured the performance of the multicasting operations (PVM's original multicast operation and our re-implementation) by iteratively executing the multicast operation. During each iteration, a timer is started, the sender then performs the multicast operation and then waits to receive positive acknowledgments from all the members of the multicast receiving group. Once all acknowledgments have been received, the timer is stopped, and another iteration begins.

In Figure 3.3, the top (bottom) graphs depict the time to perform the multicasting operation as a function of message size using the pre-existing PVM multicasting (our re-implementation).

When using PVM's existing multicasting facilities, for message sizes of 64 KB, the time to multicast to 1, 2, 3, 4, 5 remote hosts is approximately 108, 130, 184, 235, 290 milliseconds, respectively. For message sizes of 1 MB, the time to multicast to 1, 2, 3, 4, 5 remote hosts is approximately 1550, 1800, 2650, 3200, 3850 milliseconds, respectively. When using our PVM-ATM re-implementation (bottom two graphs of Figure 3.3), we observe that when increasing the number of remote hosts of the receiving pool from 1 to 4 the largest time difference is approximately 20 milliseconds. To compare, the two implementations of the multicasting operation, we derived the following approximate gain factor:

$$gain = (70 + 37n)/(95.83 + 4.17n), \tag{3.1}$$

where n is the size of the receiving pool. The numerator (denominator) was derived by examining the increase in latency caused by the increase in the number of receiving hosts when using the original PVM multicast operation (our re-implementation). At a fixed message size, the average increase per additional receiving host was used to extrapolate to the case of n receiving hosts.

Thus our re-implementation results in a performance gain (for 4 remote receiving hosts) of a factor of approximately two. For 10 remote hosts, our re-implementation results in a performance gain of a factor of approximately 3.2. Note that these performance gains are amortized by other PVM processing functions which occur during the PVM multicasting operation.

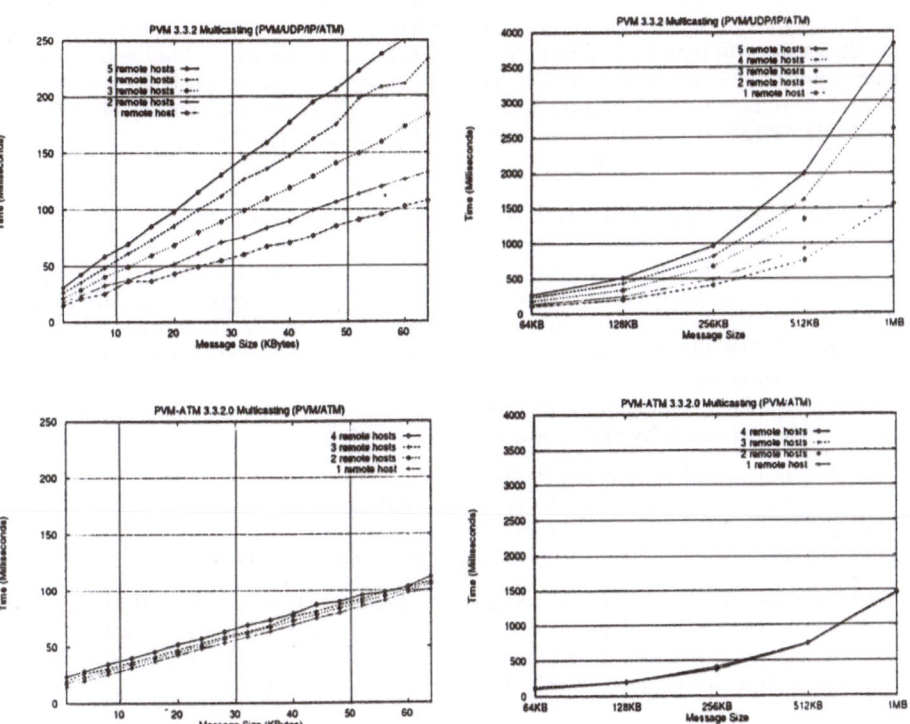

Figure 3.3. Top graphs: original PVM multicasting; Bottom graphs: re-implementation PVM-ATM multicasting

4. A Re-implementation of PVM on a HIPPI Network

The prototype of another re-implementation of PVM Version 3.3.4 (PVM/LLA) over a HIPPI LAN is presented in this section. As shown in Section 1. (Figure 1.1), the communication subsystem of PVM was originally designed to use the BSD socket interface which is a common interface for accessing standard network protocols and inter-process communications. To reduce the overhead of protocol processing and utilize the throughput of underlying networks, we re-implemented part of PVM's communication subsystem using a low-level LLA HIPPI API.

The HIgh-Performance Parallel Interface (HIPPI) [1] is one of the high-speed network or channel solutions currently commercial available. HIPPI is a simplex point-to-point interface for transferring data at peak data rates of 800 or 1600 Mbits/sec over distance of up to 25 meters. A related standard defines the usage of a cross-point switch to support multiple interconnections between HIPPI interfaces on different hosts (HIPPI-SC). HIPPI is a mature technology, most supercomputers and many high-end workstations are equipped with HIPPI interfaces for high-throughput data connections. The success and widespread use of HIPPI is due to its "KISS" (Keep It Sweet and Simple) design [17].

4.1 The LLA Application Programming Interface

The Link Level Access (LLA) application programming interface is a low-level communication interface provided in Hewlett Packard's workstation platform. The LLA interface allows application to encapsulate data into 802.2 frames. LLA uses standard HP-UX file system calls, such as *open()*, *close()*, *read()*, *write()*, *select()*, and *ioctl()*, to access the device drivers that control the network interface card. To communicate with remote processes through LLA interface, the following information must be provided by the sending task:

- SSAP: Source Service Access Point.
- Local Address: The MAC address of the sending host.
- DSAP: Destination Service Access Point.
- Destination Address: The MAC address of the receiving host.

These four tuples (*Local Address, SSAP, Destination Address, DSAP*) are used in a similar way as TCP or UDP connections. In BSD socket interface, each TCP or UDP connection is identified by {*source IP address, source port number, destination IP address, destination port number*}.

4.2 Enhanced Communications of PVM with LLA API

The LLA provides a generic communication interface for upper layer protocols (in our case, PVM's communication subsystem) to access network devices. The re-implementation of PVM over LLA can be used over Ethernet and HIPPI without any change. Upper layer processes can simply specify the device name to use any network interface and its device driver. For example, /dev/hippi represents the HIPPI interface card and its device driver, and /dev/lan0 corresponds to the Ethernet interface and driver. However, the LLA interface may have slightly different functionalities depending on the device driver of network interfaces. For example, the LLA interface provided by Ethernet doesn't support flow control mechanism, while HIPPI's LLA interface provides in-bound flow control to prevent buffer overflow at the receiving side.

The PVM/LLA is modified from the original PVM with enhancement of its communication interface and reduction of the complexity of message handling. To enhance PVM's performance, we re-implemented part of the communication sub-system of PVM using the LLA interface as follow:

1. The connections between PVM daemons was changed from UDP sockets to LLA interface. The master daemon also uses LLA to exchange messages with the shadow daemon.
2. The direct communications between local tasks and remote tasks (Direct Routing mode) is changed from TCP/IP to LLA.

As mentioned in Section 3.2, the Direct Routing mode employs TCP sockets for reliable communication. To achieve the same reliable communication as TCP sockets, the PVM/LLA relies on the sequence number of messages and the inbound flow control of LLA interface. The inbound flow control of LLA interface is inherited from the underlying HIPPI network. HIPPI provides connection-oriented service and reliable communication between hosts. The reliable communication provided by HIPPI is based on its credit-based flow control at the physical layer. The credit mechanism provides positive flow control to prevent buffer overflow at the receiving-side.

4.3 Performance Evaluation

We present the performance evaluation of a prototype PVM/LLA implementation in this Section. The experimental environment is first described in Section 4.3.1. The performance evaluation of PVM/LLA over a HIPPI network is presented in Section 4.3.2 followed by a performance tuning in Section 4.3.3.

4.3.1 Experimental Environment.

We used two HP 9000 Series 735/125 workstations which were connected with point-to-point HIPPI links. Each 735/125 workstation is equipped with 125 MHz PA-RISC processor and 80 MBytes memory. These workstations are faster than those used in the preliminary performance tests.

The HIPPI interface card is directly connected to HP's Standard Graphics Connection (SGC) I/O bus as the main I/O sub-system card. The SGC I/O bus is a 32-bit wide I/O bus which was optimized for write operations (graphical display involves lots of write operations). The theoretical throughput of the SGC bus is 60 MB/sec for outbound writing and 38 MB/sec for inbound reading. A recent performance measurement shows that the the LLA interface provided by the HIPPI interface card can achieve up to 55 MB/sec throughput for outbound transmission [15]. However, the LLA interface can only achieve around 24 MB/sec throughput for inbound reception[1].

4.3.2 PVM and PVM/LLA on HIPPI.

A set of experiments was conducted on two HP 9000 Series 735/125 workstations which were connected with point-to-point HIPPI links. In this section, the re-implemented PVM/LLA uses the LLA interface provided by the HIPPI device driver. Figure 4.1 depicts the round-trip latency measurement of the original PVM and the re-implemented PVM/LLA over the HIPPI network. For the Direct Routing mode, the re-implemented PVM/LLA shows consistent improvement of the round-trip latency. The improvement was reflected in Figure 4.1 for a wide range of message sizes, from 4 bytes to 8 Kbytes. For the message sizes shown in Figure 4.1, the re-implemented PVM/LLA achieved up to 33% of latency reduction.

---

[1] The measurement was done with a HIPPI analyzer and the *netperf* benchmark program.

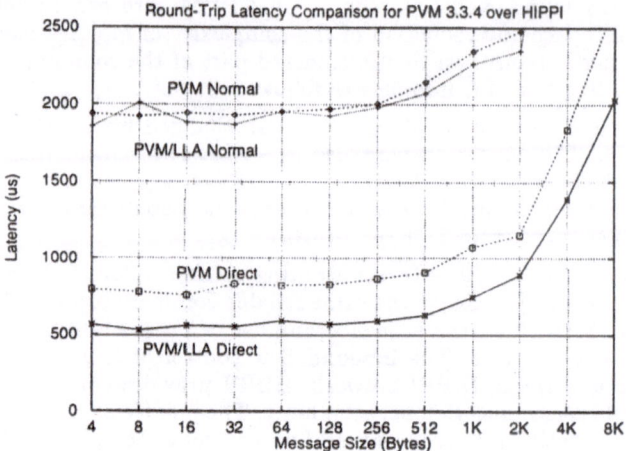

Figure 4.1. Latency measurement of original PVM and PVM/LLA over a HIPPI network.

Figure 4.2 depicts the measurement of achievable throughput of the original PVM and the re-implemented PVM/LLA over the HIPPI network. For this test, we tested five cases: (1) original PVM Normal mode, (2) original PVM Direct mode, (3) PVM/LLA Normal mode, (4) PVM/LLA Direct mode without flow control, and (5) PVM/LLA Direct mode with flow control. As shown in Figure 4.2, the re-implemented PVM/LLA can achieve higher throughput than original PVM for messages size less than or equal to 256 Kbytes. However, the achievable throughput of PVM/LLA reaches the peak with 64 Kbytes and can not obtain higher throughput after that. On the other hand, the original PVM's Direct mode reaches its peak achievable throughput 9.679 Mbytes/sec with message size of 256 Kbytes. Please note that the curve of "PVM/LLA Direct" overlaps with curve of "PVM/LLA Direct with flow control", and the curve of "PVM/LLA Normal" overlaps with curve of "PVM Normal".

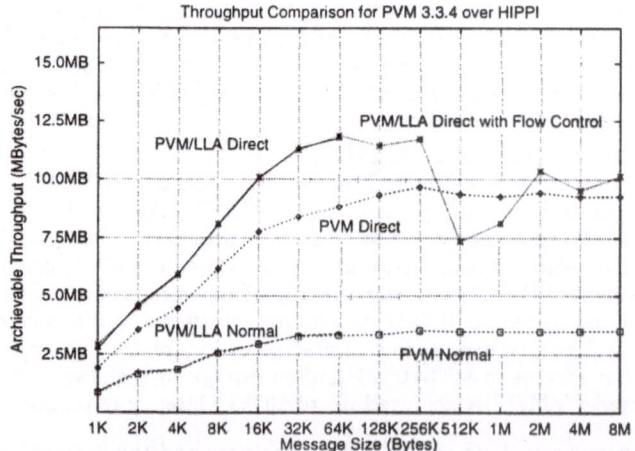

Figure 4.2. Throughput measurement of original PVM and PVM/LLA over a HIPPI network.

Figure 4.2 shows one interesting behavior of the re-implemented PVM/LLA. For message sizes larger than 256 Kbytes, the achievable throughput drops dramatically from 11.72 Mbytes/sec to 7.35 Mbytes/sec. The reason for the performance degradation is due to the throughput mismatch between the speed of HIPPI and the speed of I/O bus which can not transfer messages in a comparable speed as HIPPI. We experienced data loss problem with messages larger than 256 Kbytes. In next Section, we will demonstrate the effect of flow control and the size of message transmission unit on the performance of PVM/LLA.

4.3.3 Performance Tuning of PVM/LLA over HIPPI.

The throughput mismatch problem between the HIPPI network and the SGC I/O bus of HP Series 735/125 workstations suggests that SGC I/O bus is the bottleneck. As mentioned before, the SGC I/O bus can sustain 55 Mbytes/sec for write operations and only 24 Mbytes/sec for read operations, which are much lower than the 100 Mbytes/sec bandwidth of HIPPI network (with 32-bit data channel). To improve the performance of PVM/LLA, we should try to retrieve messages across SGC I/O bus as fast as possible. Therefore, the main design principle of our PVM/LLA is to preserve the low user-to-user latency while increasing the achievable throughput for larger messages.

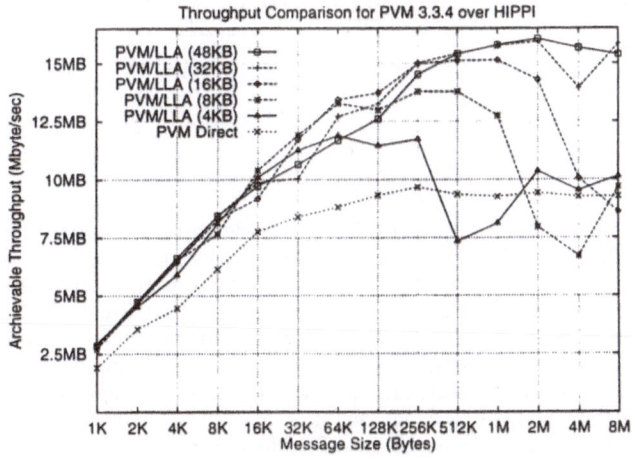

Figure 4.3. Throughput measurement of PVM/LLA with different transmission sizes.

Our early re-implementation of PVM/LLA sends out a message to the HIPPI device or receives a message from the HIPPI device using packets of 4 Kbytes, which is the default size of the transmission unit. For example, a message of 256 Kbytes will be chopped into 64 packets of 4 Kbytes for transmission. A simple optimization for improving the achievable throughput is to increase the size of packets used to transfer data between PVM tasks and the HIPPI device. It will speed up the transmission of data across the SGC I/O bus because of the reduction of the overhead from the per-packet processing. This approach is similar to the solution mentioned in Section 2. used to improve the Direct Memory Access (DMA) performance of a network I/O subsystem. In Section 2., we found that the per-page processing is the biggest bottleneck of the DMA operations. They increased the page size for each DMA operation to reduce the total overhead.

Figure 4.3 depicts the achievable throughput of PVM/LLA over HIPPI network with different transmission sizes. In this test, we increased the number of read buffers and used the flow control mechanism provided by the HIPPI LLA interface. Two interesting observations can be found in Figure 4.3. First, the peak achievable throughput of PVM/LLA was increased with larger transmission sizes. The peak achievable throughput is 11.763 Mbytes/sec with transmission size of 4 Kbytes. With transmission size of 48 Kbytes, PVM/LLA has peak achievable throughput of 16.103 Mbytes/sec. Second, the performance degradation problem was alleviated with larger transmission sizes and the activation of in-bound flow control. There is a small performance degradation when we used 32 Kbytes as the transmission size. For PVM/LLA with transmission size of 48 Kbytes, there is no performance degradation when the size of messages are less than 4 Mbytes. Table 4.1 summarizes the performance of PVM and PVM/LLA over 100 Mbytes/sec HIPPI network with the same performance metrics as before.

Table 4.1. End-to-end performance over of PVM and PVM/LLA over HIPPI.

PVM Environment	Normal Routing mode			Direct Routing mode		
	r_{max} Mbytes/sec	$n_{1/2}$ Bytes	t_0 μsec	r_{max} Mbytes/sec	$n_{1/2}$ Bytes	t_0 μsec
PVM 3.3.4	3.506	3086	1922	9.679	4717	758
un-tuned PVM/LLA	3.390	1989	1855	11.763	4050	528
tuned PVM/LLA	3.390	1989	1855	16.103	7551	540

5. Conclusions

The computing power of a single workstation and Personal Computer (PC) is increasing at a very fast pace. How to connect several workstations or PCs together to form a cluster to perform computing intensive jobs becomes an interesting research topic. The key issue of creating a high-performance workstation or PC cluster is to finding ways to reduce user level communication latency and to increase user level communication throughput. In this paper we presented some research results. Our work has been concentrated on standard switch-based high-speed networks like Fibre Channel, HIPPI and ATM. However, it is possible to use other types of interconnect like SCI (Scalable Coherent Interface [8]), ServerNet [16] and MyriNet. In fact, SCI, ServerNet and MyriNet may provide better performance than HIPPI, Fibre Channel and ATM. However, they are not yet as popular as HIPPI, Fibre Channel and ATM. We are currently investigating these configurations. Another emerging standard which may have profound impact on I/O performance is "Intelligent I/O" (I2O). This standard is still under development. It may potentially can reduce the communication latency and increase the throughput by using an extra IOP (I/O Processor).

Reference

1. ANSI X3.183-1991 (HIPPI-PH), X3.210-199x (HIPPI-FP), X3.218-199x (HIPPI-LE) and X3.222-199x (HIPPI-SC). *"High-Performance Parallel Interface"*. American National Standard Institute, Inc., June 1991, Feb 1992, June 1992, and July 1992.

2. Beguelin, A., Dongarra, J., Geist, A., Manchek,R., Sunderam, V., "A User's Guide to PVM (Parallel Virtual Machine)", Technical Report ORNL/TM-11826, Oak Ridge National Laboratory, Oak Ridge, TN, July 1991.
3. Bulter, R. and Lusk, E., "User's Guide to the P4 Programming System", Technical Report ANL-92/17, Argonne National Laboratory, 1992.
4. Carriero, N. and Gelernter, D., "Linda in Context", *Communications of the ACM*, 32(4), pp. 444-458, April 1989.
5. Geist, G.A., Sunderam, V.S., "Network-Based Concurrent Computing on the PVM System", *Concurrency: Practice and Experience*, 4 (4):293-311, June 1992.
6. Haas, Z., Protocol Structure for High-Speed Communication over Broadband ISDN, *IEEE Networks*, 5(1):64–70, January 1991.
7. Elisabeth Wechsler, *"Industry Harnesses Clustered Workstations To Squeeze Extra Cycles"*. NAS News, Vol 2, No. 9, March/April 1995.
8. IEEE Computer Society, *"IEEE Standard for Scalable Coherent Interface (SCI)"*, 1993.
9. Kolawa, A., "The Express Programming Environment", *Workshop on Heterogeneous Network-Based Concurrent Computing"*, Tallahassee, Oct. 1991.
10. Lin, M., Hsieh, J., Du, D., Thomas, J., MacDonald, J., "Distributed Network Computing Over Local ATM Networks", *IEEE Journal on Selected Areas in Communications: Special Issue on ATM LAN Implementation and Experiences with an Emerging Technology*, Vol 13, No. 4, May 1995.
11. Manchek, R.J., "Design and Implementation of PVM Version 3", *Master Thesis*, University of Tennessee, Knoxville, May, 1994.
12. Message Passing Interface Forum, *"MPI: A Message-Passing Interface Standard"*, Version 1.1: June, 1995.
13. W. Peterson. *The VMEbus handbook*. VFEA International Trade Association, 3rd Edition, 1993.
14. J. Renwick and A. Nicholson. *"IP and ARP on HIPPI"*, *RFC 1374*, Oct. 1992.
15. Rami el Sebeiti, Hewlett-Packard France, Personal communications, 1995.
16. ServerNet, Enterprise Systems Journal, June 1995, Tandem Computer Inc.
17. D. Tolmie and J. Renwick. *"HIPPI: Simplicity Yields Success"*. *IEEE Network*, page 28, Jan. 1993.
18. Von Eicken, T., Culler, T.E., Goldstein, S.C., and Schauser, K.E., Active Messages: a Mechanism for Integrated Communication and Computation, *The 19th Annual International Symposium on Computer Architecture*, pages 256–266, May 1992.
19. Zitterbart, M., High-Speed Transport Components, *IEEE Network*, 5(1):54–63, January 1991.

ATM High Performance Distributed Computing Laboratory

HPDC Laboratory Director

Salim Hariri (Email: hariri@cat.syr.edu)

Researchers

Xue Bing, Tai-Tin Chen, Wojtek Furmanski, Kiran Ganesh
Dongmin Kim, Yoonhee Kim, Joohan Lee, Vinod V. Menon
Sung-Yong Park, Ilkyeun Ra, Shaleen Saxena, Pradeep Selvakumar
Rajesh Srinivasaraghavan, Haluk Topcuoglu, Wang Wei
Baoqing Ye, Luying Zhou

CASE Center
Syracuse University
Syracuse, NY 13244
URL: http://www.atm.syr.edu

Abstract

The New York State Center for Advanced Technology in Computer Applications and Software Engineering (CASE) at Syracuse University has established an ATM High Performance Distributed Computing (HPDC) Laboratory within the center that provides ATM connections to several other units both on and off campus. This HPDC Laboratory has been established to provide a research testbed for multimedia and ATM related studies and provide a controlled environment where enabling technologies and applications can be developed and evaluated. In this paper, we describe the computing environment of HPDC Laboratory and overview some of the ongoing projects.

1 Introduction

The advent of high speed networking technology such as Asynchronous Transfer Mode (ATM) has created great opportunities and challenging research problems. ATM provides the technology necessary for full multimedia applications, which include real-time video conferencing, full-motion video and Video-on-Demand (VOD), as well as High Performance Distributed Computing (HPDC) applications.

The New York State Center for Advanced Technology in Computer Applications and Software Engineering (CASE) at Syracuse University has established a HPDC Laboratory within the center that provides ATM connections to several other units both on and off campus. The HPDC Laboratory has been constructed to provide a controlled environment where enabling technologies and applications can be developed and evaluated. The resources at HPDC Laboratory will allow experimentation with real-world applications on todays high speed networks in a manner that takes full advantage of network capabilities. Currently, there are several ongoing projects for multimedia and ATM related studies at the HPDC Laboratory. These include a High Performance Computing and Communication (NET-HPCC) design tool, a high speed message passing system for distributed computing (NYNET Communication System), Expert system based Network Management System (ExNet), and designing a Virtual Distributed Computing Environment (VDCE). These projects were selected because they represent a wide variety of network-based applications that can utilize efficiently the resources of the HPDC Laboratory. The first section of this paper describes the current environment and system architecture of the HPDC Laboratory. The second section provides an overview of the ongoing projects. Finally, the third section contains the summary and concluding remarks.

2 Network Topology at the HPDC Laboratory

The HPDC Laboratory at CASE Center at Syracuse University is based on two IBM 8260 Hubs, a GTE SPaNet ATM switch, and a Cabletron MMAC-Plus Enterprise switch. The four switches are connected using a ring topology. The Cabletron MMAC-Plus switch provides a connectivity to a Fore ASX-200 ATM switch which is located at the Northeast Parallel Architectures Center (NPAC) at Syracuse University. The Fore ASX-200 switch at NPAC provides a connectivity to the NYNET. NYNET is a high-speed fiber-optic communications network linking multiple computing, communications, and research facilities in New York State. (See Figure 1).

NYNET ATM testbed uses high-speed ATM switches interconnected by fiber-optic Synchronous Optical Network (SONET) links to integrate the parallel computers and supercomputers available at NYNET sites into one virtual computing environment. Most of the wide area portion of the NYNET operates at speed OC-48 (2.4 Gbps) while each site is connected with two OC-3 (155 Mbps). The upstate to downstate connection is through DS-3 (45 Mbps) link.

The full topology at the HPDC Laboratory is shown in Figure 2. Currently, one IBM hub has eight IBM RS6000 workstations with Turboways 155 ATM adapters, while the other has four Pentium 166 PCs with 155 Mbps OC-3 interfaces. Two of them have the PCI based ATM adapters from Zeitnet Inc and the other two have the PCI based ATM adapters from Efficient Inc. These PCs are running both Linux and Windows NT operating system. All workstations and PCs are connected using UNI 3.1 and Classical IP over ATM. Four SUN

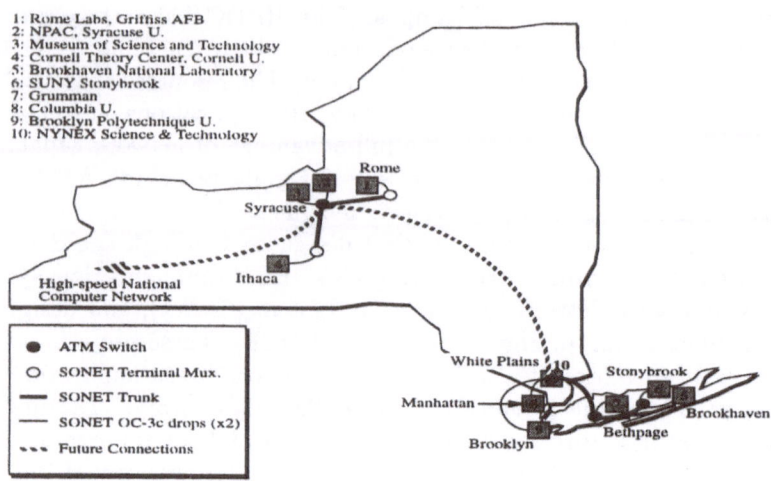

Figure 1: NYNET ATM Network Testbed

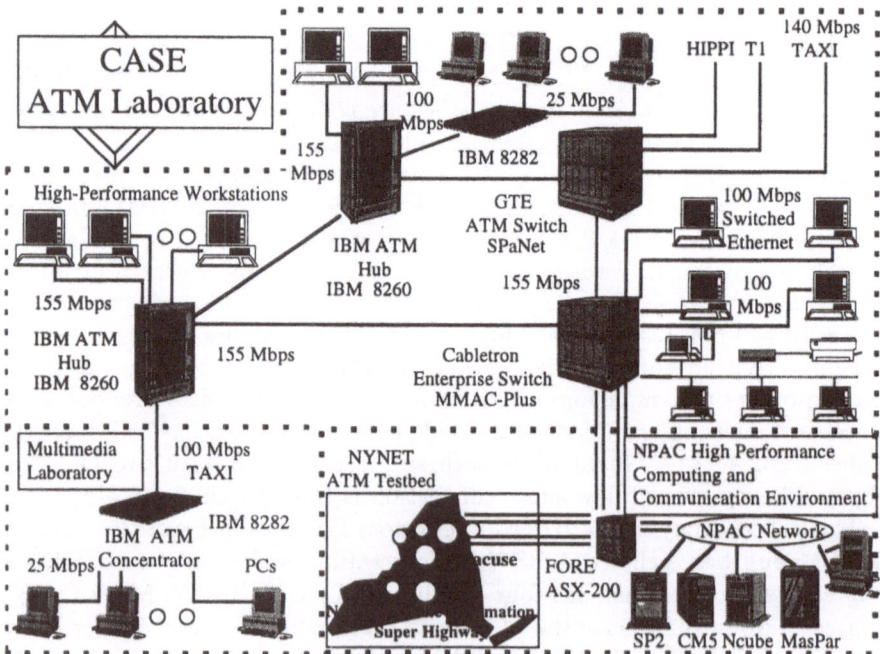

Figure 2: HPDC Laboratory Topology

SPARC-4 workstations are connected to the Cabletron MMAC-Plus Enterprise switch. They are equipped with 155 Mbps ATM adapters from SUN and connected to the MMAC-Plus using UNI 3.1 and Classical IP over ATM. The local network also provides connections from the IBM hubs to three IBM 8282 concentrators and one IBM 8285 concentrator. The concentrators provide 25 Mbps connections for PCs. These are concentrated and connected to the IBM hub through a 100 Mbps TAXI interface for IBM 8282 or a 155 Mbps OC-3 interface for IBM 8285. In addition to the OC-3 and 25 Mbps ATM connections, the workstations and PCs are connected to a 10 Mbps Ethernet LAN.

3 Current Projects

3.1 NET-HPCC: A Design Toolkit for HPCC Systems and Applications

URL: http://www.atm.syr.edu/projects/nethpcc/index.html

NET-HPCC is a research toolkit for designing High Performance Computing and Communication (HPCC) systems and applications. This toolkit uses a framework to integrate various modeling techniques and provides the HPCC system designer with an in-depth analysis of the potential performance of the system. It provides a graphical interface to the HPCC system designer, with icons that represent workstations, parallel computers, rings, buses, switches etc. The designer need not be an expert in any modeling discipline, nor does he/she need to know the specifications of various computing machines, network media and protocols. One merely needs to specify what devices/machines are present in the system and how they are connected. The toolkit provides a capability for interactive analysis of potential system designs, ruling out trouble-prone configurations and identifying potential bottlenecks. It analyzes an existing or proposed HPCC system in terms of its basic performance characteristics (throughput, utilization, latency and packet delay times). It performs sensitivity analysis of the output metric based on changes in the workload parameters, system topology and connectivity. This tool is implemented in Java, which is an object oriented language optimized for Internet applications. Applications written in Java are easily available to anyone having access to Internet through a Java compatible browser.

3.1.1 General Architecture and Analysis Approach

The design process of an HPCC system and application can be described in terms of four phases (See Figure 3): 1) System Description Phase that allows users to define the topology, network components (switches, rings, bridges, etc.), and platforms of any HPCC system; 2) Protocol Description Phase that allows users to define the type of communication protocol running on each HPCC component in the system; 3) Application Description Phase that allows users to specify the type of applications or workloads running on each HPCC component in the system; and 4) Integrated Analysis Phase that allows the users to analyze the performance of HPCC system and application and perform what-if scenario analysis using the appropriate mode of analysis.

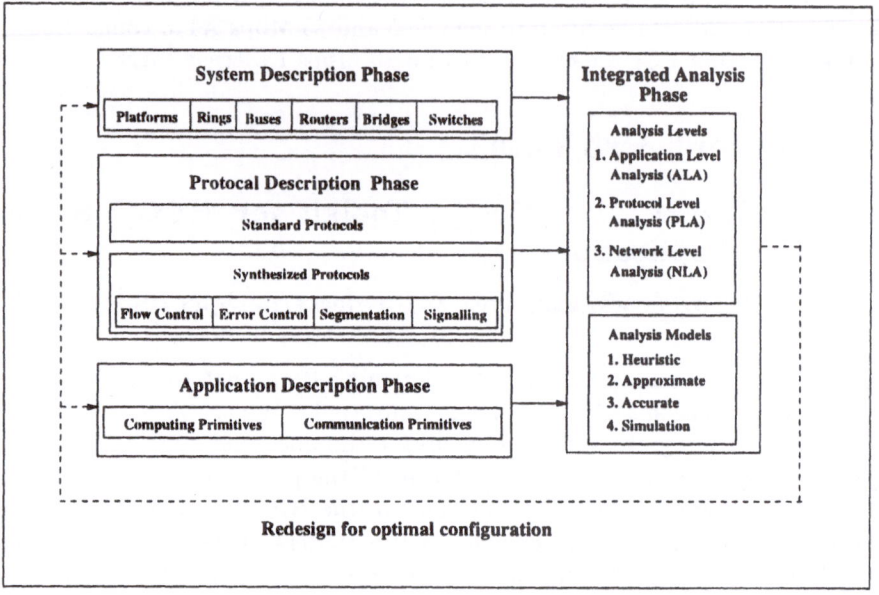

Figure 3: Phases in HPCC System Design Cycle

Based on this design process, NET-HPCC tool will have four software modules:

1. **The System Module:** The System Module consists of a library of a large number of HPCC system components. We have identified the key components to be used to build large scale high performance networks, and will also be developing analytical models for these components, such as (1) buses; (2) internetworking components; (3) switches; (4) platforms; (5) servers; (6) communication links; (7) subnets and so on.

2. **The Protocol Module:** This module allows the user to describe the type of protocol running on each network and platform components in the designed HPCC system. We offer the users two kinds of capabilities: standard protocols (e.g., TCP/IP, UDP/IP, IP/ATM, Fiber Channel, etc.) at each level of the communication software hierarchy or synthesize a new protocol where the user specifies the implementation of each protocol mechanism or function. The generic protocol development suite consists

of drag-and-drop protocol function primitives, which are classified into two types: data level parameters such as data stream segmentation, error control, flow control, routing, and session level parameters such as signaling, priority and preemption, connection establishment and termination. Using these primitives, the user can develop a protocol which is highly optimized for a given class or a set of applications (workload). Furthermore, the use of the synthesized approach will enable the users to study the performance impact of one protocol function implementation on the overall system performance.

3. **The Application Module:**   The application module is responsible for abstracting the application description into a set of primitives which defines its structure and performance. In addition, this module provides the means to describe workload characteristics running on each system component. The application module is composed of two components: (1) Machine Independent Abstraction Module (MIAM) and (2) Machine Specific Filter (MSF). In the MIAM, the user can completely characterize an application by representing it in an application graph. The primitives for building an application system are available in the application module. The primitives available are broadly classified into computing and communication elements. The computing primitives are also divided into low level primitives (such as arithmetic operations, control operations, I/O operations, communication operations etc.) and high-level primitives (such as video file operations, compression algorithms etc.). The application graph is then passed through the machine specific filter by the toolkit to incorporate machine specific information. Consequently, the analysis module can accurately analyze the application performance at the application level analysis.

4. **Integrated Analysis Module:**   In NET-HPCC, we adopt a multilevel analysis approach: 1)Application Level Analysis (ALA), 2) Protocol Level Analysis (PLA), and 3) Network Level Analysis. In this layered analysis approach, the performance of each layer is formulated independent of the other layers. The goal of such a three-pronged analysis approach is to perform the analysis in a more efficient manner, omitting all the unnecessary details, while at the same time capturing all the essential functions of the network entity. A designer can plug and unplug components at a particular level, without modifying anything at the other levels (we just use their equivalent queuing systems). This leads to a what-if analysis of the whole system with focused interest to the performance issues of the studied level. What if the protocol alone is changed at this node ? What if the underlying hardware is replaced with another one ?

3.1.2 Implementation Approach

The NET-HPCC toolkit is implemented as a Java applet that can run using a Java compatible browser (like Netscape or Microsoft Internet Explorer). Since most web browsers disallow file access on the local system, all the file I/Os are done on the host from which the toolkit is downloaded. The applet connects to a server running on the host which takes care of all the file I/Os. The server in turn connects to an mSQL database server which contains all the data for the different components that are needed for the analysis of the HPCC system (See Figure 4).

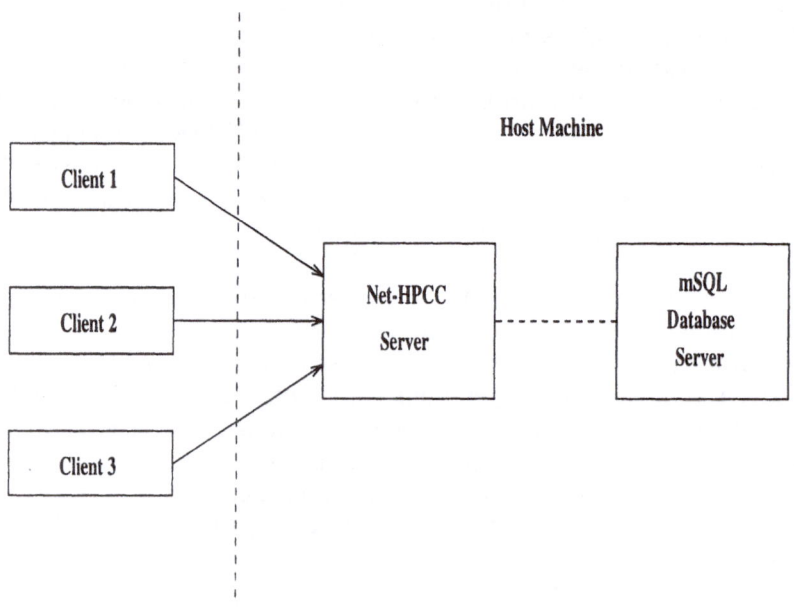

Figure 4: Net-HPCC Database Server

The server consists of a GUI for managing the database and monitoring the system access. Java was selected as the language of implementation for two reasons: It is Object Oriented and ideally suited for the development of such a complex software, especially with its highly developed windowing system (The Abstract Window Toolkit). It is Internet-friendly and hence lets the end user use the toolkit remotely through standard web browsers.

Each user will be given an account to store all the designs and the results of the analysis. The software architecture of the NET-HPCC toolkit comprises of the main NET-HPCC class which displays the buttons for the various windows that the user might want to use. Each time the user wants to use a window a separate thread is created for that window, and further instantiations of that window are not possible until the current instantiation is destroyed. The

NET-HPCC class creates objects of SystemThread, ProtocolThread, Application-Thread, MapThread and AnalysisThread classes which are all derived from the Thread class defined by Java (Figure 5).

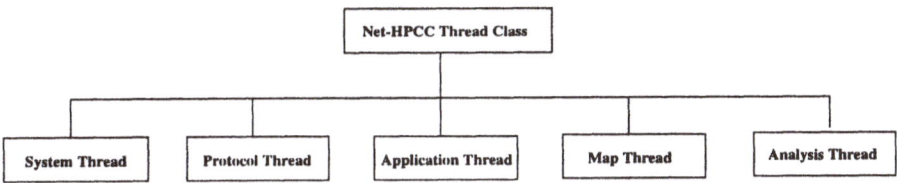

Figure 5: Net-HPCC Class Hierarchy

3.1.3   Using the Toolkit

The toolkit can be accessed using a Java compatible web browser. When it is accessed, the toolkit displays a series of buttons: System, Protocol, Application, Analysis.

When the System button is clicked, the toolkit pops up the System window. Using the menu, a network can be designed by clicking on the appropriate network and topology components (See Figure 6).

The components can be dragged and dropped around the system window. The system will disallow connecting incompatible components. By double-clicking on each of the components, the user can bring up a dialog box that specifies the various attributes of that component. The default data shown in each of the dialog boxes will be retrieved from the database. The user can also save and load systems that have been designed before and stored on the host by clicking on the file menu.

The protocol window allows the designer to specify the protocol to be used with each HPCC component of the system developed in the System window. The user selects a protocol from the menu and then clicks on the component using that protocol. By double clicking on each of the components, the user can set the various protocol parameters. The user has the option of either selecting a pre-defined protocol such as TCP/IP, or defining a new protocol in terms of protocol mechanisms such as flow control methods, error control methods etc.

The application window is then activated by selecting the Application button. Here the user specifies the kind of applications that will run on each system component.

Double clicking on the applications pops up a dialog box that takes the various attributes for that application. The user can save/load designed applications by using the file menu. There is also an option for modeling an application as a task graph, where the user decomposes an application into communication and computing elements. The computing class consists of primitives for program

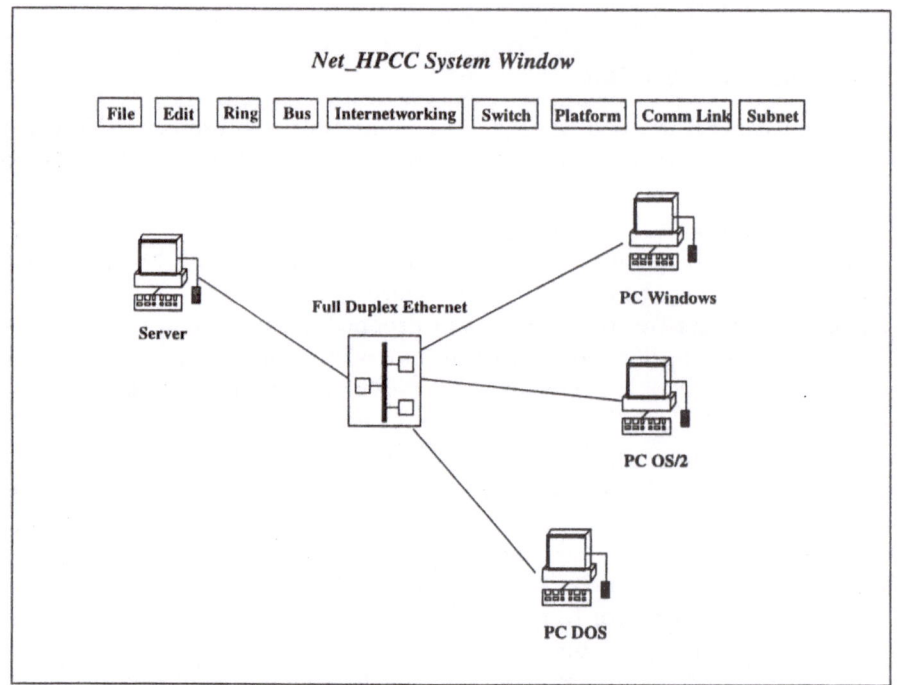

Figure 6: NET-HPCC System Window

control, floating point operations, file I/O, etc., while the communication class consists of primitives for sending and receiving data.

On clicking the Analyze button, the results of the analysis are shown such as response time, throughput and utilization times. Various modifications can be made on the different windows and the analysis window will reflect the changes (See Figure 7).

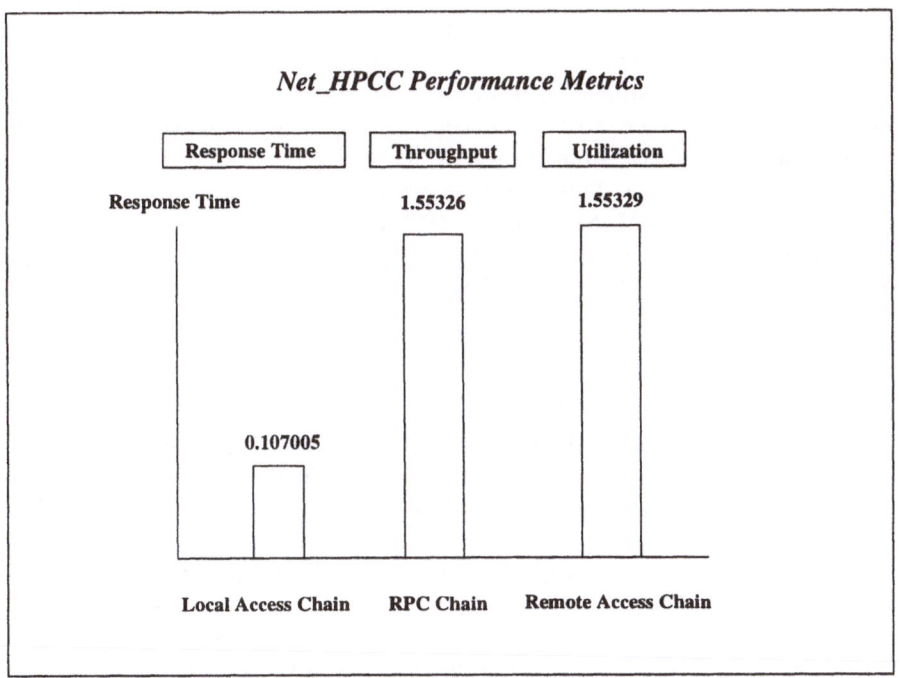

Figure 7: Net-HPCC Analysis Window

3.2 NYNET Communication System (NCS)

URL: http://www.atm.syr.edu/projects/NCS/index.html

Advanced processors and High Performance Networks, such as ATM have made high performance network computing an attractive computing environment for large-scale HPDC applications. However, the communications overhead has eliminated any performance gains provided by the technology advances. To provide a high performance communication system, we implemented a low-latency, high-throughput message passing tool which supports a wide range of HPDC applications with different Quality of Services (QOS) requirements. We call it the NYNET Communication System (NCS). The main features of

NCS include thread based programming paradigm, separation of data and control paths, overlapping of computation and communication, and high speed trap interfaces. The thread based programming paradigm allows us to overlap computation and communication and thus can reduce the propagation delay impact on performance in wide area distributed computing environment. By separating data and control paths, NCS can support several different multicast and flow control algorithms which can be activated at runtime to satisfy the multicast, flow control requirements of a given HPDC applications. This allows NCS to support efficiently a wide range of HPDC applications, where each requires different mechanisms to implement multicast, congestion control, and flow control services. NCS uses read/write trap routines to bypass traditional operating system calls. It reduces latency and avoids using traditional communication protocols (e.g., TCP/IP). This was done using the user trap routines and modified Fore Systems' ATM API on Sun IPX workstations. Our benchmarking results showed that the performance of our implementation is more than twice better than that of Fore Systems' ATM API.

3.2.1   General Architecture

In the NCS environment, a process consists of user threads and system threads. User threads are written by the user to implement the computations of a given HPDC application (we refer to them as compute threads). System threads may be control, data transfer, flow control, or multicast threads. They are created on initialization and activated by the control thread according to a user specification. Consequently, our approach allows the HPDC programmer to select optimal flow control and multicast algorithms for a given HPDC application. For example, in an ATM network, the control thread establishes ATM connections for data transfer and maintains configuration data of each machine. The data transfer threads (send and receive threads) are spawned by the control thread to perform only data transfer. Users can specify flow control (e.g., rate, credit, or window based) and multicast (e.g., repetitive send/receive or a multicast spanning tree) algorithms that are best for their applications.

Figure 8 illustrates the general architecture of NCS and shows the data connection and control connection, respectively. The NCS architecture supports process grouping to achieve scalability and improve performance (especially when NCS is applied to large scale HPDC applications). Currently only static process grouping is allowed in NCS. The future implementation will include dynamic process grouping to support applications that require dynamic grouping (e.g., Video-conferencing). Grouping is done on the basis of processes, not machines. Within each group, there is a single group server which handles all intergroup and multicast communications. Additional details on the NCS architecture can be found in [3] and [4].

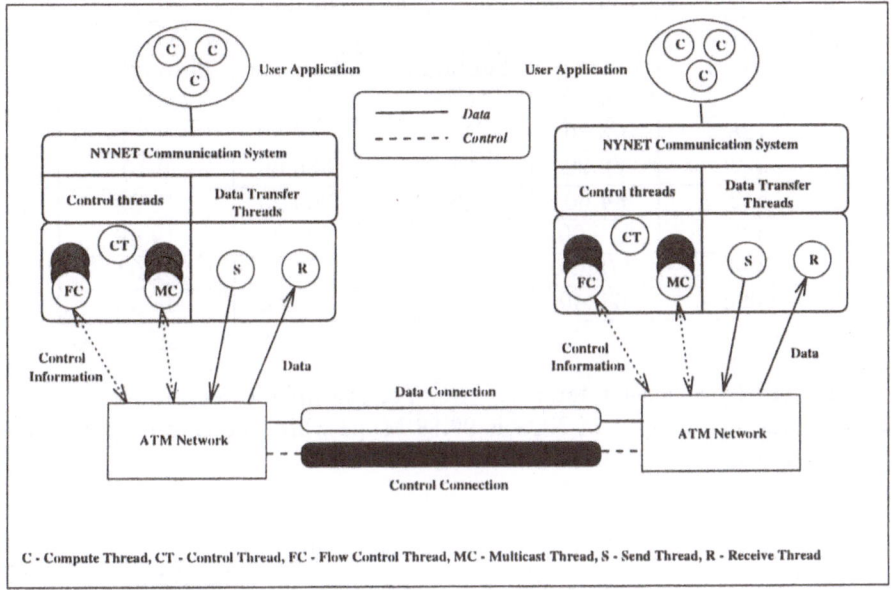

Figure 8: General Architecture of NCS

3.2.2 Benchmarking Results

The NCS is implemented over three operating modes such as Normal TCP/IP mode, Native ATM Interface mode, and High Speed Interface mode. The Normal TCP/IP mode is built upon the standard UNIX socket interface. The Native ATM Interface mode is built upon the native ATM API. The High Speed Interface mode avoids traditional communication protocols (e.g., TCP/IP) and uses a modified ATM API. In this operating mode, we reduce the overhead associated with traditional read/write system calls by allowing NCS to invoke directly the trap routines.

As of this writing, NCS send/receive using trap routines and NCS TCP/IP mode have been implemented. In what follows, we provide the benchmarking results for send latency/throughput, receive latency/throughput, and application performance using four workstations over TCP/IP operating mode, respectively.

Table 1 compares the performance of NCS send routine with that of Fore Systems' API. The message size varies from 128 bytes up to 4 Kbytes because Fore Systems' API has a maximum transfer unit of 4 Kbytes. It is clear from the Table 1 that NCS send routine outperforms Fore Systems' API by at least a factor of two. For example, given a message size of 4 Kbytes, the throughput of NCS is 101.87 Mbps while the throughput of Fore Systems' API is 45.55 Mbps.

Table 2 compares the performance of NCS receive routine with that of Fore Systems' API. Like the performance at the send side, NCS outperforms Fore

Table 1: Send Latency and Throughput

Message size	NCS		FORE API	
(bytes)	Latency (usec)	Throughput (Mbps)	Latency (usec)	Throughput (Mbps)
128	63.20	16.20	317.20	3.23
256	72.20	28.37	335.60	6.10
512	84.60	48.42	369.20	11.09
1024	119.00	68.84	429.40	19.08
2048	178.04	90.82	527.80	31.04
4096	320.40	101.87	716.60	45.55

Systems' API by at least a factor of two. For example, given a message size of 4 Kbytes, the throughput of NCS is 99.89 Mbps while the throughput of Fore Systems' API is 47.07 Mbps.

Table 2: Receive Latency and Throughput

Message size	NCS		FORE API	
(bytes)	Latency (usec)	Throughput (Mbps)	Latency (usec)	Throughput (Mbps)
128	79.25	12.92	207.75	4.93
256	87.50	23.41	216.75	9.45
512	101.00	40.55	235.75	17.37
1024	135.50	60.46	269.50	30.39
2048	198.75	82.44	347.25	47.18
4096	326.75	99.89	693.50	47.07

Table 3 compares the performance of NCS using four distributed computing applications such as Matrix Multiplication, Joint Photographic Experts Group (JPEG) Compression/Decompression, Fast Fourier Transform (FFT), and Parallel Sorting with Regular Sampling (PSRS) against MPI, p4, PVM implementations. As we can see from Table 3, the NCS implementations outperform other implementations using MPI, p4, and PVM. This experiment provides the information about the performance gain obtained from NCS by using multi-threading and separation of data and control activities. Since the flow control is not implemented at the time of this writing, we used Normal TCP/IP operating mode for our benchmarking. We believe that the NCS implementations using Native ATM Interface mode and High Speed Interface mode outperforms the Normal TCP/IP mode implementation.

Table 3: Application Performance using 4 Node - TCP/IP mode (seconds)

Applications	NCS	MPI	P4	PVM
Matrix (256x256)	10.872	15.360	21.995	25.949
JPEG (500x500)	4.528	6.054	13.351	9.126
FFT (1024 points)	1.029	1.407	1.766	1.735
Sorting (50000 items)	21.48	30.74	59.64	53.94

3.2.3 Future Related Activities

The current NCS implementation is built using the QuickThreads toolkit [6] developed at the University of Washington. Since the Quickthreads toolkit does not support variety of computing platforms, we are planning to evaluate another thread package (Pthread user-level package [7]) which is more portable and standard. We are also working on evaluating and implementing different multicast and flow control algorithms so that we can run NCS applications over ATM API directly.

Some research efforts are also undergoing to evaluate the performance of different ATM APIs as well as ATM adapter cards. This activity helps us to characterize the different parameters affecting the performance of the ATM API and ATM adapter cards. The ATM adapter cards which will be evaluated are 155 Mbps ATM cards from SUN (Solaris 2.5), 155 Mbps ATM cards from FORE Systems (SUN OS 4.1.3), 155 Mbps and 25 Mbps ATM cards from IBM (AIX 4.1 and Windows NT 3.51), 155 Mbps ATM cards from Efficient Inc. (Linux and Windows NT 3.51), and 155 Mbps ATM cards from Zeitnet Inc. (Linux and Windows NT 3.51).

3.3   Virtual Distributed Computing Environment (VDCE)

URL: http://www.atm.syr.edu/projects/vm/index.html

The continuous growth of the available power of today's computers, combined with the growing popularity of Internet and the emergent broadband network based National Information Infrastructure (NII), open new exciting opportunities but also pose new software engineering challenges for the high performance distributed computing environments.

By analogy with the rapid onset of the World-Wide-Web based global information infrastructure, we envision intuitive, user-friendly new generation tools for HPDC, scalable from LANs to WANs to World-Wide heterogeneous networks including clusters of PCs, workstations and specialized supercomputers. However, the experience from the previous attempts at building such systems indicates that the underlying software engineering problems are inherently hard and only limited successes for specialized computational domains were reported so far in this area.

We propose here a novel approach to HPDC software, based on high level abstract virtual machine concepts. Our VDCE decomposes the complex problem of building user friendly robust high performance computing environment into a set of smaller, independent but interacting layers. We try to build VDCE in a staged fashion according to the following three-prong approach: analysis of the existing enabling technologies and partial solutions; conceptual full system design; early prototyping, driven by one selected application domain (military command & control). We believe that this strategy will allow us to address the 'in large' software engineering challenges while staying in touch with the previous and ongoing related efforts, and refining the overall design based on lessons learned in focused prototyping activities.

We adopted visual hierarchical data-flow paradigm as the top level user metaphor. Hence, for end-users, VDCE offers intuitive authoring tools which allow to build distributed applications by composing computational graphs in terms of library modules, represented as icons and easily accessible in the click-and-drag or click-to-connect mode. Such computational graphs are then automatically mapped by the middleware subsystem on the available computing resources, and finally executed in a controlled fashion by a Virtual Machine software subsystem in the backend.

Several functions involved in the VDCE model outlined above such as visual graph authoring, scheduling, load balancing, fault tolerance or performance monitoring were previously addressed by various HPDC systems. Our goal here is to integrate these concepts, algorithms and strategies in the form of a full system design, and to develop a proof-of-the-concept prototype.

3.3.1 Architecture Overview

The VDCE architecture outlined above can be abstracted in terms of the following three major software layers: Application Development Layer (ADL) which offers visual graph authoring tools and generates computational graphs that specify the computation (modules/nodes) and communication (links) of a distributed application. Application Configuration Layer (ACL) which maps such task graphs, optionally augmented by user hints and preferences on the available system resources; Application Execution Layer (AEL) which receives task-resource mapping information and constructs an application specific Data Virtual Machine (DVM) to be executed under control of an application-independent Control Virtual Machine (CVM). In the following, we briefly characterize the individual VDCE layers listed above.

Application Development Layer (ADL)

Visual graph based representation for distributed applications is a popular technique, successfully exploited in scientific visualization packages such as Application Visualization System (AVS) [16] or Khoros [17], HPCC toolkits such as CODE [12] or HeNCE [15], and most recently also in Internet authoring tools such as ActiveX by Microsoft. Typical approach for a given computational domain is to accumulate a set of core library modules that encapsulate the base functionality of a given domain, and then develop composite modules and/or full applications using suitable graphical composition tools.

For the VDCE testbed, we intend to use as our library base a set of Command, Control, Communication and Information (C3I) functions, specified recently in Rome Lab C3I Parallel Benchmarking project [8]. These include modules such as Hypothesis Testing, Plot-Track Assignment, Route Optimization, Terrain Masking, Threat Analysis, Decision Support System etc. Using this base, one can then start composing various customized C3I/Battle Management subsystems such as the one depicted in Figure 9.

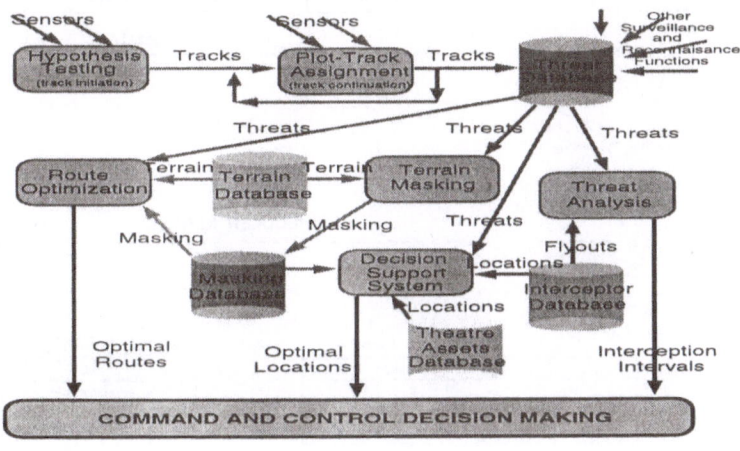

Figure 9: Example of visual computational graph for C3I/Battle Management, constructed in terms of Rome Lab C3I Parallel Benchmarking modules/library functions

Application Configuration Layer (ACL)

The task graph specified by the ADL is passed to the ACL where it is mapped on the available system resources. Each node of a VDCE system maintains the local resource database with the machine and network configuration parameters such as: memory size, processor type, attached devices, operating system,

CPU, security constraints, reservation policies, network latency and bandwidth, current and time-averages load, etc. These local databases, augmented by the networking and connectivity parameters, form effectively a system resources graph that represents the actual heterogeneous hardware of a VDCE system.

The goal of the scheduler unit in ACL is to find the optimal match between system resources graph and the task flow graph given by ACL. End-users might require fully automated, autonomous scheduling, whereas more sophisticated users might want to provide various hints and scheduling preferences which need to be taken into account by the scheduling algorithms.

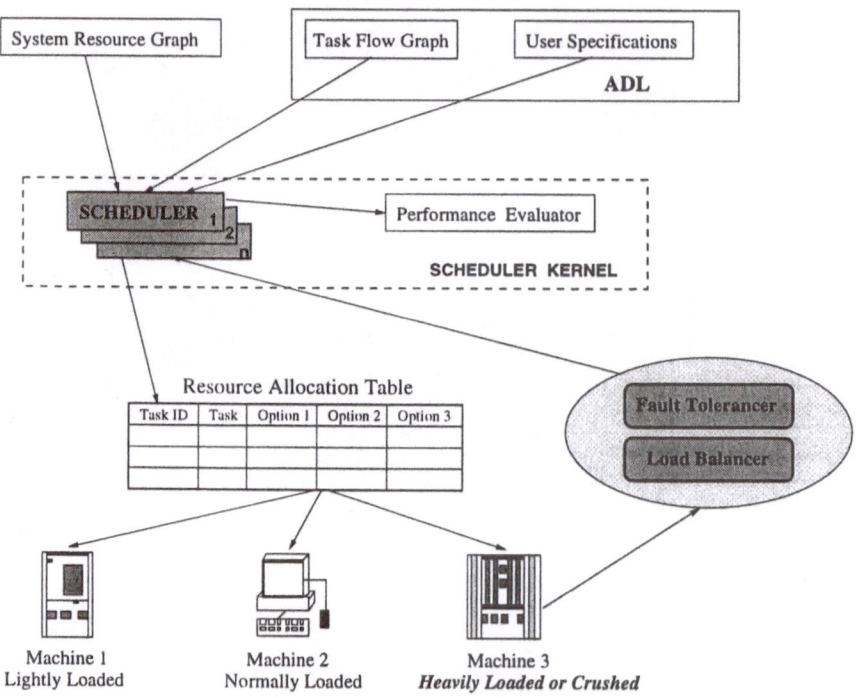

Figure 10: Application Configuration Layer

Scheduling is in general an NP-hard combinatorial problem and we therefore intend to provide a set of alternative algorithms and strategies, to be selected depending on the problem at hand and user preferences. Usually, a set of sub-optimal solutions will be generated and maintained as a list of scheduling options. The best solution is executed when the application starts and the other options can become useful and instantiated later on in case of faults, load fluctuations or other runtime changes in the system resources graph.

Application Execution Layer (AEL)

Task-Resource Mapping Table generated by ADL is passed to the AEL for execution. AEL itself splits into two components: a CVM, given by a set of daemons continuously running in all VDCE nodes, and a DVM, constructed by CVM for the purpose of the optimal execution of the specific application, provided by the ADL and ACL subsystems.

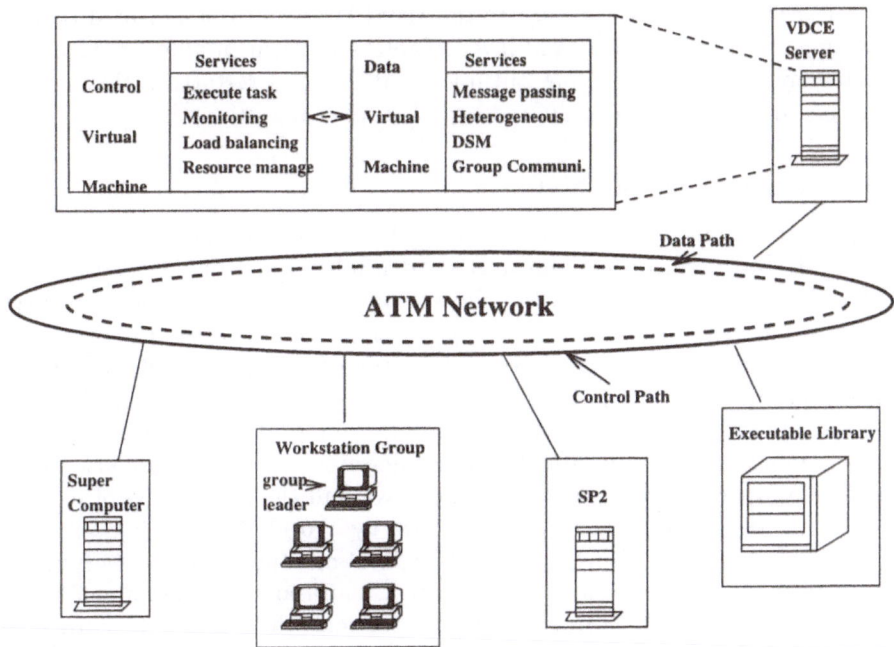

Figure 11: Application Execution Layer

We plan to support two communication topologies for DVM: high performance message passing and shared distributed memory. We intend to use multi-threaded point-to-point message passing packages such as Nexus [14], Chant [13] or NCS [3], and to evaluate a set of available Distributed Shared Memory (DSM) packages such as Treadmarks [9], CRL [10] or Adsmith [11].

The process of building a DVM for a given application involves the following steps: construct executables for individual modules by linking the abstract VDCE message passing calls, say *vdce_send()* and *vdce_receive()*, present in pre-compiled modules with the suitable high performance message passing libraries; establish socket-based point-to-point communication channels between modules; establish connections between DVM modules and their CVM proxies to be used for performance monitoring and visual feedback purposes. Once a DVM is constructed and the execution starts, CVM switches to the monitoring mode to assure smooth, fault tolerant and performance optimized (load balanced) execution. A keep alive packet is circulated along a virtual token ring circuit linking all DVM daemons to detect faulty nodes and rearrange the network if necessary. The same or similar circuit is used to propagate and accumulate the current and average load information. Realtime data collected in these monitoring activities provides continuous feedback to the ACL layer which is ready to dynamically adjust new DVM configuration in case of faults of load changes. Such reconfiguration decisions are passed back to the DVM layer and implemented there by adjusting or rebuilding the DVM layer. Typical adjustment amounts to moving selected modules to new more optimal locations and restarting their actions there based on the periodically updated checkpointing information.

3.3.2 Implementation Status

Early prototype of some aspects of VDCE has been constructed recently for a matrix algebra application (Linear Equation Solver) and demonstrated at the HPDC-5 conference in August '96. We now initiate a more complete prototype development with the focus on C3I application domain discussed above.

Visual graph editing tools are being constructed using Java and GEF (Graph Editing Framework) [20] package. We are also analyzing the design of CODE package and intend to reuse some concepts for our ADL and ACL layers.

In parallel with these frontend and middleware activities, we are also initiating the CVM prototyping activities using Java and the new family of dynamic Java based Web servers such as Jigsaw [18] by World Wide Web Consortium (W3C) or Jeeves (Java Web Server) [19] by JavaSoft. Individual CVM functions identified so far such as Querier, Dispatcher, Creator and Destructor can be naturally mapped on URL-addressable Jeeves servlets which facilitate testing and fine-tuning individual CVM services during this layer development. Proxies of DVM modules will be also constructed as Java objects, dynamically instantiated in CVM servers and Java socket-connected with their DVM computational counterparts.

3.3.3 Summary

We presented an overview of the current design and prototyping activities of the VDCE system. We are currently starting the prototype development for the C3I application domain and we expect early demos by February '96. Lessons learned in this prototyping period will provide feedback to the full system design prong which is expected to terminate by summer '97 and to deliver the consistent and tested VDCE design, supported by proof-of-concept prototypes and ready for production level implementation in the anticipated next stage of the project (fall '97).

3.4 Intelligent Network Management System (ExNET)

URL: http://www.atm.syr.edu/projects/ENM/enms1.html

In this project, we are evaluating the use of rule-based and case based reasoning techniques in the management of large scale high speed networks in order to develop an intelligent network management system prototype (ExNet). ExNet modules will be incorporated with commercially available network management systems (e.g. IBM NetView). The architecture of the ExNet System is detailed in the next section.

3.4.1 ExNet Architecture

ExNet System consists of four main modules : Monitor, Expert System, Network Interface and Network Manager Interface modules. The general architecture is shown in Figure 12.

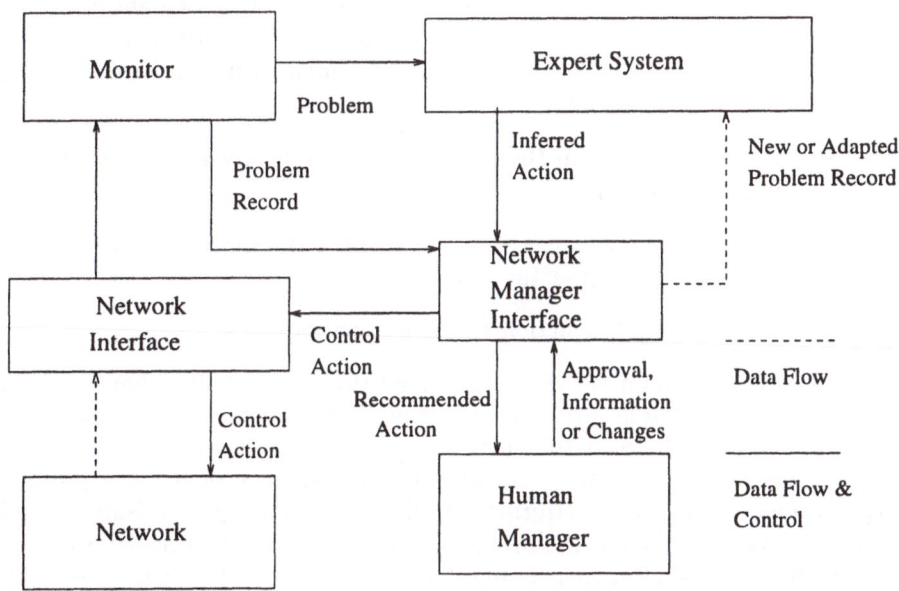

Figure 12: Generic Expert System for Network Management

The Monitor would be involved in monitoring the general health of the network. It would keep track of the changes that are occurring within the network. Furthermore if it deems any event to be critical to the performance of the network it will report it to the expert system module. The Expert System would make the appropriate analysis of the situation provided to it by the monitor. Then based on its internal reasoning it will decide a specific course of action to alleviate that problem. In our project, we intend to build both rule-based and case-based reasoning expert systems. Both of these approaches have certain disadvantages and advantages. As a part of our effort we would like to compare both these design methodologies. The Network Interface will be responsible for transferring of information about events as they happen in the network. Also it will be responsible for the implementation of strategies recommended by the expert system, and approved by the human manager. The Network Manager Interface will be the contact point between the human administrator and the ExNet system. The Network Manager Interface will present the information about the network, such as the current status of various nodes and links to the manager. It will also present the solution that the expert system has come up with to the human manager.

3.4.2 Expert System Prototype

We have developed a prototype of the expert system module. It was interfaced through rudimentary monitor, network interface and network manager interface modules. We tested it for a hypothetical network consisting of eight hosts and four gateways. Our main aim was to simulate conditions for a real network and to route data between any two hosts. The main tasks performed in this implementation were as follows:

- Monitoring congestion in the network by periodically gathering information on system status.

- Deciding whether the congestion has exceeded thresholds or not and what nodes are more congested than others.

- Suggesting a corrective action to the human manager.

The data regarding the network was provided through text files that had been generated based on real network observations. The network interface was hence a simple subroutine to read text files. The monitor received data about the network from the network interface and decided whether or not the problem in the current routing was performance related or some system fault related. The expert system analyzed the problem, requested more information from the network interface and after considering the alternatives decided on the new route to be implemented between the two given hosts.

In this implementation example, we had used three set of rules: 1) Check CPU State, 2) Check-Routing-ResponseTime, 3) Diagnose-network-status. The first rule checked the loads on each computer to determine whether or not it is highly loaded. The second rule evaluates whether the response time associated with the current network route meets the desired performance requirements, so that this route can be changed when the performance requirements can not be met. The third rule shows how the system can prevent from sending new requests to highly loaded machines or network devices.

3.4.3 Web interface to NetView

In this Activity we have developed CGI scripts to perform elementary tasks of network management. We have also developed web-based interfaces to NetView functions.

1. **Basic functions:** The basic capabilities that we have are based on UNIX commands. These commands allow us to perform various activities. For example we can ping any host to check if it is alive and to determine the response time. By using the *netstat* command interface statistics on a host can be obtained. It displays the traffic in packets per second for the given host. Another command allows us to determine the current activities taking place on a host. It gives information such as the users logged on to the system and the system resources being used by each of them. Another set of utilities allow us to determine the name of the host from the IP address and vice versa. Lastly two more utilities allow us to determine the platform and the operating system of the given host.

2. **NetView Interfaces:** We have developed some scripts to run some NetView command and utilities. We use the *xnmgraph* command to graph the traffic characteristics on each of the interfaces of a selected host. The *ovobjprint* command is used to obtain system informations about a host within the network managed by NetView. The command *snmpColDump* is used to display the traffic conditions recorded by NetView on a host.

3. **Traffic Monitoring Script:** We have certain scripts that obtain and store in a text file the amount of traffic on two of our workstations. These scripts are based on the *netstat* command in UNIX. The *netstat* command returns the total amount of traffic that has passed through its interfaces since the machine was booted last. Our script runs the *netstat* command on an hourly basis. The data returned by the *netstat* command contains the total number of input octets, total number of input collisions, total number of output octets and total number of output collisions. This data is stored in a file. Since we have the total traffic scenario on an hourly basis, we can obtain the hourly traffic scenario by simply taking the difference of the consecutive terms in the first data file. The corresponding hourly traffic data is stored in a second file. On the very first data collection of the day, another section of the same program updates a link in the calendar presented on the web.

4. **NetView Programming:** In large computer networking environment, the management of computer and network resources is extremely difficult and cumbersome. Our goal is to automate the management activity by linking expert systems (eg. ExNet) to a commercial network management

system (eg. NetView). Figure 13 shows our approach to integrate these two systems.

Some of the managed activities, like load on a CPU or traffic passing through a router, may cross the threshold of safety for that parameter. In the SNMP based scenario that we are considering, the SNMP agent responsible for that particular host or node will generate a trap. This trap will be received by the NetView daemons. These daemons will trigger the Exnet System by passing the appropriate parameters concerning the event. Exnet will decide on the corrective measures to be taken. These measures when implemented will correct or reduce the damage done by the fault.

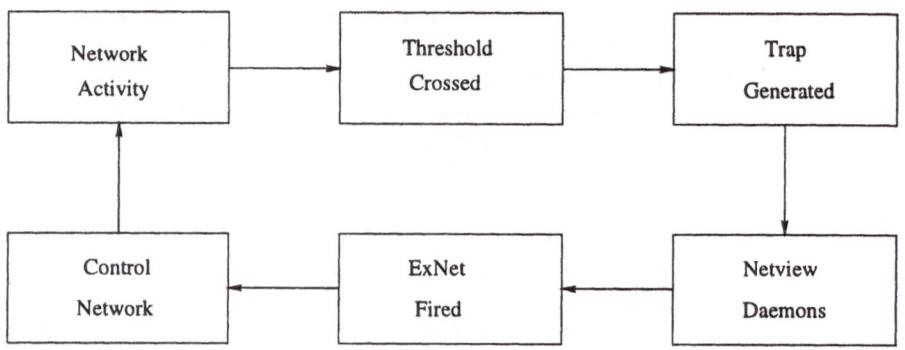

Figure 13: Loop of Events

4 Conclusion

In this paper, we reviewed some of the ongoing projects at the HPDC Laboratory. However, to efficiently utilize the emerging high performance computing and networking technologies, more research is needed in three areas: (1) high speed protocols, (2) host-to-network interfaces, and (3) parallel and distributed software tools.

1. *High Speed Protocols:* The overhead associated with standard protocols (e.g., TCP, UDP) and operating systems is high and it prevents parallel and distributed applications from achieving bandwidth comparable to the network speed (Currently, the application transfer rate is about 10 - 20 % of the network transfer rate [3].). We do need protocols that provide flexible communication services with low overhead.

2. *Host-to-Network Interface:* The host-to-network interface design plays an important role in reducing overhead and achieving high throughput. Re-

search is needed to develop techniques that reduce the overhead associated with data movement, data copying, and error checking.

3. *Parallel and Distributed Software Tools:* Developing efficient parallel and distributed computing application is a non-trivial task. It requires the users to have a good knowledge of the intrinsic characteristics of the parallel and distributed systems, networks, and tools. We do need research that produces software tools that assist the users to experiment with, debug, and evaluate different parallel and distributed algorithms running on different high performance computing configurations.

References

[1] S. Hariri et. al., "Net-HPCC, A Design Tool for High Performance Computing and Communications: Part-1, Performance Modeling", October 1996 (unpublished document)

[2] S. Hariri et. al., "Net-HPCC, A Design Tool for High Performance Computing and Communications: Part-2, Software Design", 1996 (unpublished document)

[3] S. Y. Park, S. Hariri, Y. H. Kim, J. S. Harris, and R. Yadav, "NYNET Communication System (NCS): A Multithreaded Message Passing Tool over ATM Network", Proc. of the 5th International Symposium on High Performance Distributed Computing, pp. 460-469, August 1996.

[4] S. Y. Park, I. K. Ra, M. J. Ng, S. M. Park, and S. Hariri, "An Efficient Multithreaded Message Passing Tool for ATM-based Distributed Computing Environments", Proc. of the 9th International Conference on Parallel and Distributed Computing Systems, September 1996.

[5] S. Hariri, S. Y. Park, R. Reddy, M. Subramanyan, R. Yadav, and M. Parashar, "Software Tool Evaluation Methodology", Proc. of the 15th International Conference on Distributed Computing Systems, pp. 3-10, May 1995.

[6] D. Keppel, "Tools and Techniques for Building Fast Portable Threads Packages", University of Washington, Technical Report UWCSE 93-05-06.

[7] F. Mueller, "A Library Implementation of POSIX Threads under UNIX", Proc. of USENIX Conference Winter '93, pp. 29-41, January 1993.

[8] S. Hariri, G. Fox, B. Thiagarajan, D. Jadav, M. Parashar, "Parallel Software Benchmarking for BMC3/IS Systems", Technical Report (SCCS-490), Northeast Parallel Architectures Center (NPAC), 1994.

[9] C. Amza, A. L. Cox, S. Dwakadas, P. Keleher, H. Lu, R. Rajamony, W. Yu and W. Zwaenepoel, "TreadMarks: Shared Memory Computing on Networks of Workstations", IEEE Computer, Vol. 29, No. 2, pp.18-28, February 1996.

[10] K. L. Johnson, M. F. Kaashoek and D. A. Wallach, "CRL:High-Performance All-Software Distributed Shared Memory", SIGOPS '95, pp. 213-228, December 1995.

[11] W. Y. Liang, C. King, F. Lai, "Adsmith: An Efficient Object-based Distributed Shared Memory System on PVM", Technical Report TR-951201, Dept. of Computer Science and Information Engineering, National Taiwan University, 1995.

[12] P. Newton and J. C. Browne, "The CODE 2.0 Graphical Parallel Programming language", Proc. of ACM International Conference on Supercomputing, July 1992.

[13] M. Haines, D. Cronk, P. Mehrotra, "On the Design of Chant: A Talking Threads Package", Supercomputing'94, pp. 350-359, November 1993.

[14] I. Foster, C. Kesselman, S. Tuecke, "The Nexus Approach To Integrating Multithreading and Communication", Journal of Parallel and Distributed Computing, 1996.

[15] R. Wolski, C. Anglano, J. Schopf, F. Berman, "Developing Heterogeneous Applications Using Zoom and HeNCE", Heterogeneous Workshop, IPPS, 1995.

[16] AVS 4.0 Developer's Guide and User's Guide, May 1992.

[17] Maui High Performance Computing Center, http://www.mhpcc.edu/training/workshop/html/khoros/khoros.html, 1994.

[18] World Wide Web Consortium, http://www.w3.org/pub/WWW/Jigsaw.

[19] Java Soft, Inc. http://www.javasoft.com:80/products/java-server/index.html.

[20] Dept. of Information and Computer Science, University of California at Irvine, http://www.ics.uci.edu/ jrobbins/GraphEditingFramework.html.

Switched Virtual Networking

Detlef Straeten, Martin Mähler
IBM European Networking Center
Heidelberg

Introduction

The Asynchronous Transfer Mode (ATM) technology is frequently seen only under the aspects of bandwidth gain and the suitability for multi-media communication. An essential advantage of ATM however is its potential for new concepts of connection management in complex networks. By the ability of ATM to create virtual subnets the separation of the logical network topology from the physical location of the subnet participants becomes possible. In conjunction with other switching LAN technologies like switched Ethernet and switched Token Ring extremely flexible and dynamic communication structures can be build which reflect the organisational requirements of nowadays business processes.

In this way the concepts of switched networks and virtuality are going to become the new technical fundamentals for networking architectures. While introducing new concepts interoperability with the huge existing installed base is imperative to be successful in the marketplace. Therefore aspects of integration and migration of traditional infrastructure will be critical to any new approach.

This article describes a model how to implement these principles step by step. New functionality will be introduced into the network at those places first where today applications will benefit immediately, e.g. at the server access or backbone to eliminate bottlenecks. This way new technologies like ATM, Virtual LANs (VLANs) and MAC-Layer Switches can be integrated while simultaneously taking care about future applications that will demand end-to-end Quality of Service (QoS) connections. The article therefore starts with a short overview of existing LAN technologies, their main properties and the way to interconnect them. Based on this the Switched Virtual Networking (SVN) model [2] will be introduced. The concepts of *virtual LANs and the options how to implement them will be* described. The main options will be compared by explaining how they define the role of networking components like routers and higher layers protocols.

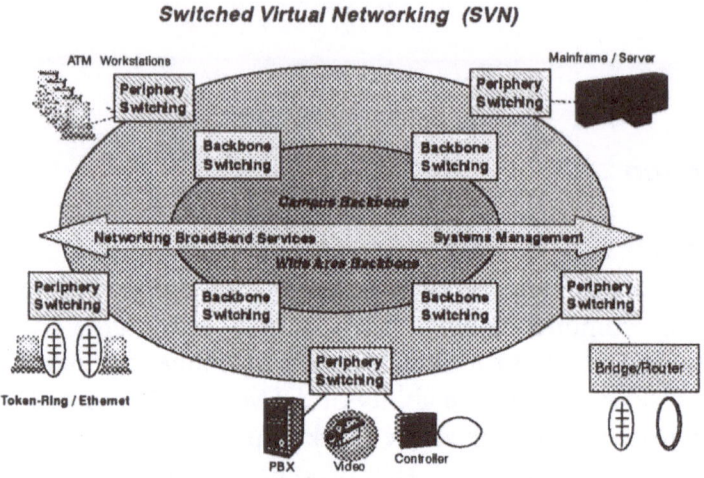

Figure 1: SVN model

Today's LANs

Today's enterprise networking structures are result of a tremendous growth process over the last decades. The introduction of LAN technology (Ethernet, Token Ring) to the enterprise infrastructure made it possible to exchange information between users with a speed not known before. Although the technology was based on a „shared medium" the resulting bandwidth per user was exiting compared to kilobit rates of former point-to-point connections. Early LANs were characterised by few segments with a small number of users though. The situation began to change when more and more users were connected to the LAN resulting in less bandwidth for each user. The only way to keep pace was to constantly increase the number of LAN segments in order to provide a certain amount of bandwidth to everyone. Moreover, new bandwidth demanding applications put additional load on the network. The resulting process of structural *change in enterprise networks is also called 'microsegmentation'. The* important consequence is that a variety of interconnection devices found to be needed in the network to keep users connected on different segments. Bridges, routers and eventually switches therefore move more and more to the focus of interest.

While bridges are sometimes called „plug and play" devices routers need to be configured according to a number of protocol specific parameters. For this complexity routers are rather „plug and call hotline" devices that require specific skill and experience. Both bridges and routers only accomplish the task to provide connectivity. It's up to network administrator to put them at the right places in order to achieve a good topology with regard to throughput and delay. In fact, running and managing a 'microsegmented' enterprise network turned out to be a very complex task. One reason for this complexity is the undeterminism of traffic characteristics and traffic flows. It's very difficult to provide a constant service level when user behaviours and demands change dynamically. For example, heavy file transfer traffic interferes with short time client/server queries on a nondeterministic basis. Another serious issue is the need of dealing with multiple different protocol layers (SNA, NetBios, TCP/IP, Novell IPX, AppleTalk, etc.) which were introduced over time into the network and have their own merits and characteristics.

The SVN Model

The SVN model picks up this situation and describes how new technologies like ATM and VLAN can be introduced in the enterprise network to overcome the current problems. Simply speaking, the SVN approach is to start with switching technology in the backbone and extend it over time to the periphery. This means that the functionality of traditional bridges and routers will be moved from the core of the network to its edges. In figure 1 the main architectural components of SVN are illustrated.

The SVN architecture consists of four main functions:

- Periphery Switching
- Backbone Switching
- Networking Broadband Services (NBBS)
- Network Management

To implement these functions all components of the network have to be taken into account. The scope of SVN products therefore ranges from adapter cards for client and server workstations over campus switches to ATM wide area switching cabinets. In the following paragraphs the main functions of SVN will be described in more detail.

Periphery Switching

Periphery Switching builds the link to connect traditional equipment and new ATM enabled devices to a high-speed backbone of a campus or wide area network. For example, this means that periphery switches map traditional Token Ring and Ethernet interfaces to ATM. Moreover, periphery switches provide all standard bridging (e.g., source route, transparent) and routing (e.g., OSPF, RIP) protocols to support existing bridge/router based installations.

By using switches in the periphery and the backbone all connected terminals appear to be direct neighbours to each other. One consequence is the move of routing decisions from the core of the network to the edges, i.e. deciding in periphery switches. The resulting scenario of a so called Virtual Router allows to establish QoS connections end-to-end since there is no more router device in the data path. The concept of virtual routing is described later in a special section.

Backbone Switching

The core network infrastructure in SVN is build on ATM backbone switches. Backbone switching transports both the productive data traffic and the signalling information, e.g. needed to establish a virtual LAN or a Virtual Router. To provide the needed high level of availability and efficiency ATM technology is the right choice. ATM backbone switches support the standards specified by the ATM Forum and the ITU. In those cases where the standardisation committees currently not define interfaces SVN implements own concepts in order to provide the needed functionality. However, by observing and contributing to the standardisation process (UNI, P-NNI, MPOA, etc.) first implementations are as close as possible to the evolving standards (see also figure 6). The goal is to reduces the migration effort to comply to the final standards to a minimum.

Networking BroadBand Services (NBBS)

To run a high-speed network that meets the requirements of corporate customers a control architecture is needed that picks up the standards and closes the existing gaps. NBBS is SVN's control architecture to run ATM networks effectively and provide a variety of interfaces to it [1]. NBBS was originally designed according to the special needs of wide area networks. NBBS' initial focus therefore was on access, transport and network control services. In the context of SVN the NBBS concept now is extended by components to provide multiprotocol switched services (MSS) as well. In the following the components of NBBS shall be introduced.

NBBS Access Services provide a number of standard interfaces like Frame Relay or ISDN to connect equipment to a high-speed network that does not have ATM capability natively. This is very important in order to protect the existing investments and allow the coexistence of traditional and new technology in the same network. Without access services as a central implementation approach a smooth migration to a switched infrastructure will not be possible.

NBBS Access Services support point-to-point but also point-to-multipoint communications which become important for several scenarios, e.g. distribution of multimedia information.

NBBS Transport and Control Services are of special meaning in the wide area. Since WAN connections typically are below 2 Mbit/s an efficient bandwidth management is key to achieve QoS requirements like end-to-end delay in voice communications. Since different traffic streams (voice, data, multimedia) cause different demands NBBS provides support for a variety of traffic descriptors (e.g. peak, mean, burstiness) and mechanisms to meet them simultaneously. The mechanisms used are compliant with ATM Forum defined methods (e.g. rate based control, leaky bucket) but extend them in several areas like dynamic bandwidth adaptation and equivalent capacity concepts [1].

NBBS Multiprotocol Switched Services (MSS)

Simply speaking, MSS provides the services within a switched infrastructure which overcome the weaknesses of existing router based networks like performance bottlenecks and lacking QoS guarantees. MSS therefore contains the following components:

- Virtual LAN support
- Broadcast Management
- Enhanced ATM LAN Emulation
- Distributed Routing

MSS can be seen as another access service for the multiprotocol (TCP/IP, IPX, SNA, etc.) environment which is based on standards defined by the ATM Forum or the Internet Engineering Taskforce(IETF). See also figure 2 for an overview of supported functions.

Multiprotocol Switched Services

- **Distributed Routing**
  - Removes the router from the data path
  - Distributes layer 3 routing function to the network periphery
  - Seamless migration path for existing routers
    - Switching infrastructure appears as a router to attached existing routers
  - Provides routing between virtual LANs, between Classic IP and LAN emulation
  - Future proofing for Multiprotocol Over ATM (MPOA) standards
- **Virtual LAN**
  - Virtual LANs defined by users
  - Reduces network complexity and administration costs
- **Enhanced ATM LAN Emulation**
  - Supports larger emulated LANs
  - Multiple redundant LANE servers
  - User can be member of multiple emulated LANs
- **Broadcast Management to reduce Overhead for IP, NetBios and IPX/SAP**

Figure 2: MSS Multiprotocol Switched Services

Virtual LANs

A virtual LAN (VLAN) is defined by the set of end systems which independently of their physical location shall belong to a given LAN. The options how an end system can be assigned to a VLAN are:

- Layer 3 protocol (NetBios, IP, Appletalk, ...)
- MAC address
- Physical port

This means for example that all members of a given subnet can be grouped together as if they were connected to a physical LAN. End systems can be part of several VLANs simultaneously. If VLANs are connected (bridged) to traditional LANs the resulting network is sometimes called a virtual LAN segment or simply a broadcast domain. The aim of the VLAN concept in MSS is to group together end stations according to their communication relations rather than their physical locations. The VLAN therefore can be seen as the ideal infrastructure for workgroup communication which can be flexibly build across the existing wiring (see figure 3). The resulting network of different VLANs can be quite flat, i.e. only few router transitions are needed to communicate between them.

The following paragraphs will discuss briefly two approaches to define virtual LAN segments.

LAN Emulation

LAN Emulation, defined by the ATM Forum, defines a mechanism to emulate different MAC layers (Token Ring, Ethernet) on a common ATM transport infrastructure. Since ATM is a connection oriented protocol a client/server approach is used to emulate a traditional LAN behaviour. One task of this client/server system is address resolution of a virtual MAC address to an ATM address performed by a LAN Emulation Server. The other task is the emulation of broadcast capabilities via a multicast function called Broadcast and Unknown Server (BUS). The ATM Forum LAN Emulation specification defines the necessary interfaces [4].

In SVN the MSS server and MSS client provide and extend the ATM Forum LAN Emulation service. The according MSS server and client functions are implemented in the backbone and periphery switches, respectively. The main features are:

- Existing application interfaces in the end systems are preserved. The ATM interface appears as a Token Ring or Ethernet attachment.
- Transparent for higher layer protocols, like IP or NetBios
- Interconnection of traditional LANs and ATM end systems
- Interconnection of LAN-to-LAN via ATM segments
- LAN Emulation service redundancy
- Intelligent broadcast management

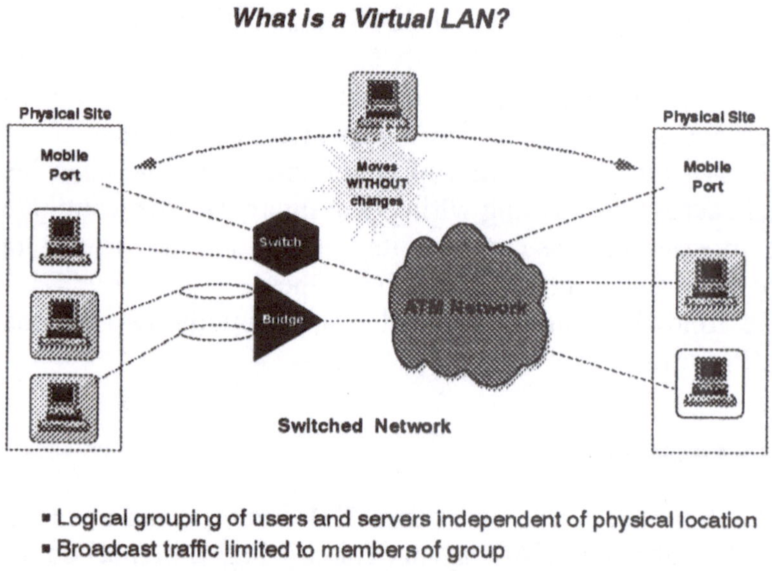

Figure 3: Virtual LANs

Classical IP

A competitive approach to transport layer 3 information across an ATM segment was defined by the IETF named 'Classical IP'. This approach defines a way to adapt the layer 3 IP protocol stack to the new layer 2 ATM base. That means it is limited to IP. Similar to the LAN Emulation approach a so called ARP-Server is used to map IP addresses to ATM addresses. The protocols and frame formats used are defined in detail in RFC1483 and RFC1577.

From a virtual LAN point of view Classical IP defines a virtual LAN composed out of all end systems participating in the same IP subnet. This is often called Logical IP Subnet (LIS). As a difference to LAN Emulation, Classical IP has no broadcast or multicast support. Protocols which use multicast or broadcast mechanisms on layer 3 (RIP, OSPF, ...) are not supported using RFC1577.

Both approaches have currently no redundancy support specified by the corresponding standardisation committees. That means, as soon as a server goes down the VLAN does not any longer exist. This is true for the LAN Emulation Server and the ARP Server what is not acceptable to productive environments. Therefore in SVN MSS Server redundancy is provided in case a LAN Emulation Service goes down. MSS Server can be duplicated inside the network so the backup MSS Server will take over in case of a failure (see figure 4).

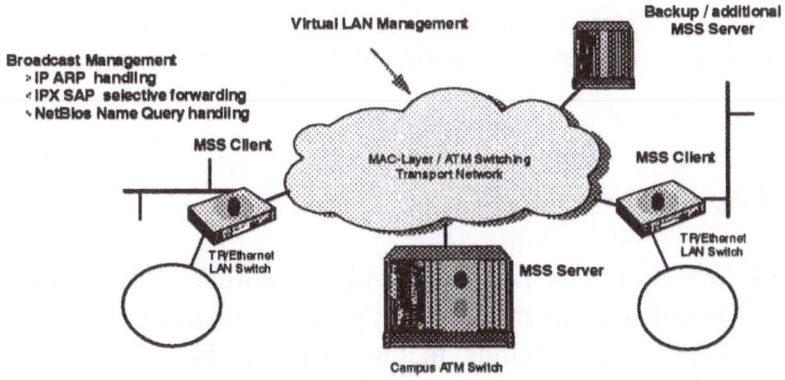

MSS LAN Emulation and Broadcast Management

Figure 4: Broadcast Manager

Broadcast Management

Based on this implementation of server instances (LAN Emulation, BUS, ...) it is possible to create large flat networks. That means, it is possible to define as much end systems per VLAN as desired. The drawback of this approach is the increasing amount of broadcasts per VLANs. This effect is known for a long time now in traditional LANs, but can be reduced by enhancing the existing BUS functionality with some learning intelligence. This functionality is called Intelligent BUS (I-BUS) or Broadcast Manager. The I-BUS monitors every broadcast and correlates MAC layer information with Layer 3/2 information elements like IP addresses or NetBios names. Based on this learning mechanism many of the broadcasts (like IP-ARP, or NetBios Name Query) can be converted into unicasts. The reduction of broadcast traffic enables network administrators to define more end systems per VLAN. This means less VLAN segments and therefore less segment interconnections. Moreover, it reduces significantly the complexity of the infrastructure. From the ATM end system point of view this I-BUS behaviour is completely transparent. This enables vendor specific implementations which distinguish each other from purely standard compliant product realisations. Figure 4 summarizes the implementation of MSS functions related to LAN Emulation and Broadcast Management.

VLAN Application

In the future Corporate Networks will use more and more switching and VLAN concepts to migrate to highspeed infrastructures. In case ATM is used, cell switching in the backbone in conjunction with LAN switching on the floor level is dominant. Server farms are natively ATM attached and assigned to a number of different VLANs. This increases the throughput for traditional attached end systems connected via LAN switches/bridges or routers. Due to the possibility to run multiple VLANs on the same ATM network the communication paths do not cross multiple hierarchy levels between source and destination. The number of bridges and routers on the path are significantly reduced compared to traditional microsegmented networks. This results into less delay and better overall performance.

Routing will still play an important role to connect different virtual or real LAN segments. The question is how many of these routing decisions (i.e. routers) are actually needed on the path between source and destination? Today's traditional approach to connect LAN segments is a 'Multi-Hop-Routing' approach. In this context the ATM network is seen as just another subnetwork where routing is performed at the ingress and egress points. In contrast, the MSS components of the SVN concept uses a distributed routing model. This approach creates a collection of components called Virtual Router which utilises the ATM network as its backplane. This is described in more detail in the next paragraph.

Distributed Routing

Another role of the MSS Server is to act as a multiprotocol router. That means it is able to make layer 3 routing decisions for those LAN MAC frames which are received. Therefore the MSS Server has a so called virtual MAC-Address which describes the entry port of the Virtual Router. As soon as a MSS Client is installed remotely on an edge device the entry of the Virtual Router moves from the integrated MSS Server in the backbone switch to the MSS Client inside the edge LAN switch. A similar discussion helds for the egress port of the Virtual Router. The ATM network becomes the backplane of the Virtual Router.

After the edge device has received a MAC frame indicating the virtual MAC address of the Virtual Router, the MSS Client will further investigate with the MSS Server what to do. The MSS Server has to be asked for a routing decision (see figure 5). The MSS Server replies with the final layer 2 destination MAC address, avoiding the traditional layer 3 next hop routing. The layer 3 routing address to layer 2 destination MAC address mapping has been previously performed by the destination MSS Client using traditional ARP caching or advanced automatic learning functions. These mappings are communicated on a regulary bases to the MSS Server. Based on the reply from the MSS Server the source MSS Client can now *switch the MAC frame in question. That means the MSS Clients are* at the same time entry and exit of the Virtual Router. Additionally the MSS Clients will cache information to avoid unnecessary server queries.

Using the above concept a so called 1-hop-routing scenario is implemented. The Multiprotocol Over ATM (MPOA) working group of the ATM Forum is discussing currently this Virtual Router approach. In this specification the components are called MPOA Server and MPOA Client. There are significant contributions to the subject of multi-protocol switching in order to standardise this kind of distributed routing. SVN's MSS Server and Client will integrate in the future the MPOA specifications as soon as they evolve.

MSS Distribited Routing

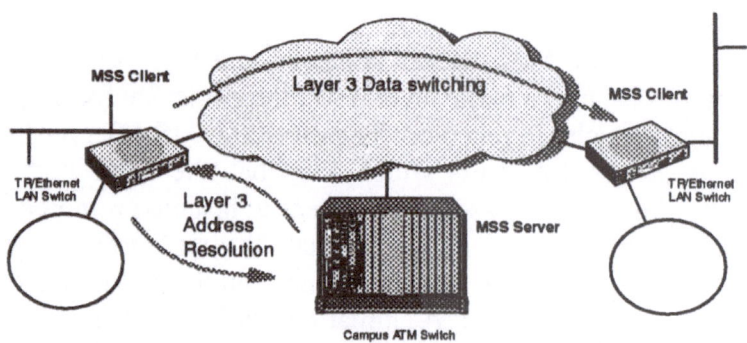

Figure 5: Distributed Routing

Using this idea of a Virtual Router one can even imagine implementing the MSS Client code on a traditional Token Ring or Ethernet LAN adapter. An application on an end system sends a higher layer protocol frame down the communication stack to communicate with a partner end system across the network. Before this frame is send on the Ethernet or Token Ring medium a MSS shim client code is investigating with the MSS Server the mapping of the destination end system address to the final destination MAC address. On getting the answer from the server the shim code on the adapter changes the destination MAC frame address and sends the resulting frame onto the Token Ring or Ethernet LAN. With this scenario a so called 0-hop-routing can be achieved. The MSS implementation in SVN is related to the following activities of the ATM Forum and the IETF. See also figure 6.

ATM Forum:

- LAN Emulation 1.0 [4]
- LAN Emulation 2.0 (work in progress), the main issue is to implement redundancy for the LAN Emulation Server components.
- MPOA Multiprotocol Over ATM [3]
- Extensions to P-NNI (Privat-Network/Network Interface (I-PNNI, PAR)) [5,6,7]. Attempt to integrate different topology views of routers and ATM switches in the same network [8].

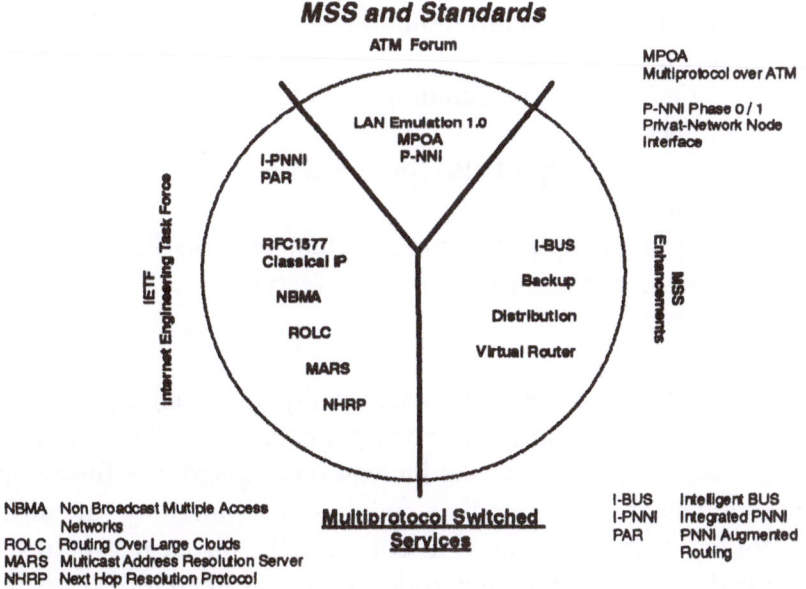

Figure 6: MSS and Standards

IETF Internet Engineering Taskforce

- MARS, Multicast Address Resolution Server. This is used to define how to implement multicasts in NBMA (non broadcast multiple access) networks, like ATM [8].
- NHRP, Next Hop Resolution Protocol. NHRP enables short-cuts between end systems attached to the same ATM network but residing on different LISs [3].

SVN Network Management

Using above complex concepts to build Corporate Networks there is a need for powerful network management systems. These systems have to support the rollout of VLAN concepts and at the same time help network managers to administer the resulting hybrid structure of traditional LANs, VLANs and routers.

Key capabilities of such management systems are:

- all components of a Switched Virtual Network can be managed
- physical views and different logical views of the network are supported
- specific VLAN configuration and VLAN management support is available
- management support also for non-SVN and multivendor environments
- management applications are implemented on different management platforms (TME 10, Netview, ...) based on standards

Today's management systems focus usually on managing the physical infrastructure, the hardware components and the connections between each other. Due to the increased complexity of future virtual networks based on cascading, mixing and merging of different protocol layers there is a need to correlate the physical structure with the logical structure of the network. Operators need tools to navigate between physical and logical views of the network. To support simple configuration management for VLANs, like

- LANs, which are built by hubs
- broadcast domains, which are built by LAN switches
- ATM LAN Emulation, Classical IP
- MSS Multiprotocol Switched Services (NHRP, MARS, MPOA,...)

there is a need for the same graphical user interfaces. Operators can transparently change the configuration using drag and drop mechanisms. Additional management disciplines like auto discovery, status monitoring, performance management, fault management and security management have to be integrated in the same platform.

The SVN management is based on standards, like Simple Network Management Protocol (SNMP) or Common Management Information Protocol (CMIP), which are specified by the ATM Forum, IETF or ITU. All equipment which is manageable based on above standards can be integrated and managed in an SVN environment.

Conclusion

Switched Virtual Networking is a key strategy to prepare a Corporate Network to fulfil the requirements of existing and future applications. Its central approach is a smooth migration to implement an end-to-end highspeed network infrastructure based on virtual LANs and distributed routing. This maximises the price/performance relationship while still supporting maximum flexibility. Existing SVN products, starting with the desktop, via Campus and WAN components to global networking infrastructures, give the ability to begin where most benefits and immediate quality improvements can be achieved. To meet the challenges of building a future proofed corporate infrastructure, SVN describes a networking model that planners can start implementing today.

References

[1] IBM International Technical Support Center, Networking Broadband Services (NBBS), Architectural Tutorial, Redbook GG24-4486

[2] IBM Technical Whitepaper, Switched Virtual Networking, 1996

[3] ATM Forum, 95-0824r8, Baseline Text for MPOA, Caralyn Brown et al.

[4] Implementing ATM Forum-compliant LAN Emulation, 1996 Digital Communications Design Conference, Bill Ellington, Cedell Alexander

[5] ATM Forum, 96-0352, The Relationship between MPOA and Integrated P-NNI, Ross Callon, et al.

[6] ATM Forum, 96-0354, An Overview of PNNI Augmented Routing, Ross Callon

[7] ATM Forum, 96-0355, Issues and Approaches for Integrated PNNI, Ross Callon

[8] ACM SIGCOMM, Computer Communication Review, Volume 25, Number 2

Parallel Large Scale Finite Element Computations

Arnd Meyer

Faculty of Mathematics, Technical University Chemnitz
D-09107 Chemnitz

1. Introduction

From Amdahl's Law, the efficient use of parallel computers can not mean a parallelization of some single steps of a larger calculation, if in the same time a relatively large amount of sequentiell work remains or if special convenient data structures for such a step have to be produced with the help of expensive communications between the processors. From this reason, our basic work on parallel solving partial differential equations was directed to investigating and developing a natural fully parallel run of a finite element computation – from parallel distribution and generating the mesh – over parallel generate / assembly step – to parallel solution of the resulting large linear systems of equation and post–processing.

So, we will define a suitable data partitioning of all large F.E. data that permits a parallel use within all steps of the calculation.

This is given in detail in the following Chapter 2. Considering a typical iteration method for solving a linear finite element system of equations, as is done in Chapter 3, we conclude that the only relevant communication technique has to be introduced within the preconditioning step. All other parts of the computation show a purely local use of private data. This is important for both message passing systems (local memory) and shared memory computers as well. The first environment clearly uses the advantage of having as less interprocessor communication as possible. But even in the shared memory environment we obtain advantages from our data distribution. Here, the use of private data within nearly all computational steps does not require any of the well–known exprehensive semaphore–like mechanisms in order to secure writing conflicts. The same concept as in the distributed memory case permits the use of the same code for both very different architectures.

2. Finite Element Computation and Data Splitting

Let

$$a(u, v) = \langle f, v \rangle \tag{2.1}$$

the underlying bilinear form belonging to a p.d.e. $\mathcal{L}u = f$ in Ω with boundary conditions as usual. Here, $u \in H^1(\Omega)$ with prescribed values on parts Γ_D of the boundary $\partial\Omega$ is the unknown solution, so (2.1) holds for all $v \in H_0^1(\Omega)$ (with zero values on Γ_D). The Finite Element Method defines an approximation u_h of piecewise polynomial functions depending on a given fine triangulation of Ω.

Let $\mathbb{V}_h$ denote this finite dimensionl subspace of finite element functions and $\mathbb{V}_{h0} = \mathbb{V}_h \cap H_0^1(\Omega)$. So,

$$a(u_h, v) = \langle f, v \rangle \quad \forall v \in \mathbb{V}_{h0}$$

is the underlying F. E. equation for defining $u_h \in \mathbb{V}_h$ (with prescribed values on Γ_D).

(In more complicated situations such as linear elasticity u is a vector function).

With the help of the finite element nodal base functions

$$\Phi = (\varphi_1, \cdots, \varphi_N)$$

we map u_h to the N-vector $\underline{u}$ by

$$u_h = \Phi\underline{u} \tag{2.2}$$

(often $\varphi_i(\mathbf{x}_j) = \delta_{ij}$ for the nodes $\mathbf{x}_j$ of the triangulation, so $\underline{u}$ contains the function values of $u_h(\mathbf{x}_j)$ at the j-th position, but it is basically the vector of coefficients of the expansion of u_h with respect to the nodal base Φ)

With (2.2) (for u_h and for arbitrary $v = \Phi\underline{v}$) (2.1) is equivalent to the linear system

$$K\underline{u} = \underline{b} \tag{2.3}$$

with

$$
\begin{aligned}
K &= (k_{ij}), & k_{ij} &= a(\varphi_j, \varphi_i) \\
\underline{b} &= (b_i) & b_i &= \langle f, \varphi_i \rangle & i, j = 1, \cdots, N.
\end{aligned}
$$

So, from the definition, we obtain 2 kinds of data:

I: large vectors containing "nodal values" (such as $\underline{u}$)

II: large vectors and matrices containing functional values such as $\underline{b}$ and K.

From the fact that these functional values are integrals over Ω, the type-II-data is splitted over some processors as partial sums, when the parallelization idea starts with domain decomposition.

That is, let

$$\overline{\Omega} = \bigcup_{s=1}^{p} \overline{\Omega}_s \quad , \quad (\Omega_s \cap \Omega_{s'} = \emptyset, \forall s \neq s')$$

be a non-overlapping subdivision of Ω . Then, the values of a local vector

$$\underline{b}_s = (b_i)_{\mathbf{x}_i \in \overline{\Omega}_s} \in \mathbb{R}^{N_s}$$

are calculated from the processor P_s running on Ω_s-data independently of all other processors and the true right hand side fulfils

$$\underline{b} = \sum_{s=1}^{p} H_s^T \underline{b}_s \qquad (2.4)$$

with a special (only theoretically existent) $(N_s \times N)$ -Boolean-connectivity matrix H_s. If the i-th node $\mathbf{x}_i$ in the global count has node number j locally in $\overline{\Omega}_s$ then $(H_s)_{ji} = 1$ (otherwise zero).

The formular (2.4) is typical for the distribution of type-II-data, for the matrix we have

$$K = \sum_{s=1}^{p} H_s^T K_s H_s \ , \qquad (2.5)$$

where K_s is the local stiffness matrix belonging to $\overline{\Omega}_s$, calculated within the usual generate/assembly step in processor P_s independently of all other processors. Note that the code running in all processors at the same time in generating and assemblying K_s is the same code as within a usual Finite Element package on a sequential one processor machine. This is an enormous advantage that relatively large amount of operations included in the element by element computation runs ideally in parallel. Even on a shared memory system, the matrices K_s are pure private data on the processor P_s and the assembly step requires no security mechanisms.

The data of type I does not fulfil such a summation formula as (2.4), here we have

$$\underline{u}_s = H_s \underline{u} \qquad (2.6)$$

which means the processor P_s stores that part of $\underline{u}$ as private data that belongs to nodes of $\overline{\Omega}_s$.

Note that some identical values belonging to "coupling nodes" $\mathbf{x}_j \in \overline{\Omega}_s \cap \overline{\Omega}_{s'}$ are stored in more than one processor. If not given beforehand such a compatibillity has to be guaranteed for type-I-data. This is the main difference to a F.E. implementation in [10], where the nodes are distributed exclusively over the processors. But from the fact that we have all boundary information of the local subdomain available in P_s , the introduction of modern hierarchical techniques (see Chapter 4) is much cheaper.

Another advantage of this distinguishing of the two data types is found in using iterative solvers for the linear system (2.3) in paying attention to (2.4), (2.5) and (2.6). Here we watch that vectors of same type are updated by vectors of same type, so again this requires no data communication over the processors. Moreover, all iterative solvers need at least one matrix–vector–multiply per step of the iteration. This is nothing else than the calculation of a vector of functional values, so it changes a type-I into a type-II-vector without any data transfer again:

Let $u_h = \Phi \underline{u}$ an arbitrary fuction in $\mathbb{V}_h$, so $\underline{u} \in \mathbb{R}^N$ an arbitrary vector, then $\underline{v} = K \underline{u}$ contains the functional values

$$v_i = a(u_h, \varphi_i) \quad i = 1, \cdots, N,$$

and

$$\underline{v} = \left(\sum H_s^T K_s H_s \right) \underline{u} = \sum H_s^T K_s \underline{u}_s = \sum H_s^T \underline{v}_s,$$

whenever $\underline{v}_s := K_s \underline{u}_s$ is done locally in processor P_s. From the same reason, the residual vector $\underline{r} = K\underline{u} - \underline{b}$ of the given linear system is calculated locally as type-II-data.

3. Data Flow within the Conjugate Gradient Method

The preconditioned conjungate gradient method (PCGM) has found to be the appropriate solver for large sparse linear systems, if a good preconditioner can be found and introduced. Let this preconditioner signed with C, the modern ideas for constructing C and the results are given in the next chapters. Then PCGM for solving $K\underline{u} = \underline{b}$ is the following algorithm.

PCGM

Start: define start vector $\underline{u}$

$$\underline{r} := K\underline{u} - \underline{b}, \quad \underline{q} := \underline{w} := C^{-1}\underline{r}, \quad \gamma_0 := \gamma := \underline{r}^T \underline{w}$$

Iteration: until stoppin criterion fulfilled do

(1) $\underline{v} := K\underline{q}$

(2) $\delta := \underline{v}^T \underline{q}, \quad \alpha := -\gamma/\delta$

(3) $\underline{u} := \underline{u} + \alpha\underline{q}$

(4) $\underline{r} := \underline{r} + \alpha\underline{v}$

(5) $\underline{w} := C^{-1}\underline{r}$

(6) $\hat{\gamma} := \underline{r}^T \underline{w}, \quad \beta := \hat{\gamma}/\gamma, \quad \gamma := \hat{\gamma}$

(7) $\underline{q} := \underline{w} + \beta\underline{q}$

Remark 3.1. The stopping criterion is often :

$$\gamma < \gamma_0 \cdot tol^2 \Rightarrow stop.$$

Here, the quantity

$$\underline{r}^T C^{-1} \underline{r} = \underline{z}^T K C^{-1} K \underline{z}$$

with the actual error $\underline{z} = \underline{u} - \underline{u}^*$ is decreased by tol^2, so the $KC^{-1}K$−Norm of $\underline{z}$ is decreased by tol.

Remark 3.2. The convergence is guaranteed if K and C are symmetric, positiv definit. The rate of convergence is linear depending on the convergence quotient

$$\eta = \frac{1 - \sqrt{\xi}}{1 + \sqrt{\xi}} \quad \text{with} \quad \xi = \lambda_{\min}(C^{-1}K)/\lambda_{\max}(C^{-1}K).$$

For the parallel use of this method, we define

	$\underline{u}, \underline{w}, \underline{q}$	to be type-I-data
and	$\underline{b}, \underline{r}, \underline{v}$	to be type-II-data
		(from the above discussions) .

So the steps (1), (3), (4) and (7) do not require any data communication and are pure arithmetical work with private data. The both inner products for δ and γ in step (2) and (6) are simple sums of local inner products over all processors:

$$\gamma = \underline{r}^T \underline{w} = \left(\sum H_s^T \underline{r}_s \right)^T \underline{w} = \sum \underline{r}_s^T H_s \underline{w} = \sum_{s=1}^{p} \underline{r}_s^T \underline{w}_s .$$

So the parallel preconditioned conjugate gradient method is the following algorithm (running locally in each processor P_s):

PPCGM

Start:

Choose $\underline{u}$,   set $\underline{u}_s = H_s \underline{u}$ in P_s

$\underline{r}_s := K_s \underline{u}_s - \underline{b}_s$, $\underline{w} := C^{-1} \underline{r}$ (with $\underline{r} = \sum H_s^T \underline{r}_s$)

set $\underline{w}_s = H_s \underline{w}$ in P_s

$\gamma_s := \underline{r}_s^T \underline{w}_s$ $\gamma := \gamma_o := \sum_{s=1}^{p} \gamma_s$

Iteration: until stopping criterion fulfilled do

(1) $\underline{v}_s := K_s \underline{q}_s$

(2) $\delta_s := \underline{v}_s^T \underline{q}_s$ $\delta := \sum_{s=1}^{p} \delta_s$, $\alpha := -\gamma/\delta$

(3) $\underline{u}_s := \underline{u}_s + \alpha \underline{q}_s$

(4) $\underline{r}_s := \underline{r}_s + \alpha \underline{v}_s$

(5) $\underline{w} := C^{-1} \underline{r}$ (with $\underline{r} = \sum H_s^T \underline{r}_s$)

     set $\underline{w}_s = H_s \underline{w}$

(6) $\gamma_s := \underline{r}_s^T \underline{w}_s$, $\hat{\gamma} := \sum_{s=1}^{p} \gamma_s$, $\beta := \hat{\gamma}/\gamma$, $\gamma := \hat{\gamma}$

(7) $\underline{q}_s := \underline{w}_s + \beta \underline{q}_s$

Remark 3.3. The connection between the subdomains Ω_s is included in step (5) only, all other steps are pure local calculations or the sum of one number over all processors.

A proper definition of the preconditioner C fulfils three requirements:

(A)   The arithmetical operations for step (5) are cheap (proportionally to the number of unknowns)

(B)   The condition number $\kappa(C^{-1}K) = \xi^{-1}$ is small, independent of the discretization parameter h (mesh spacing) or only slightly growing for $h \to 0$, such as $\mathcal{O}(|\ln h|)$.

(C)   The number of data exchanges between the processors for realizing step (5) is as small as possible (best: exactly one data exchange of values belonging to the coupling nodes).

Remark 3.4. For no preconditioning at all ($C = I$) or for the simple diagonal preconditioner ($C = diag(K)$), (A) and (C) are perfectly fulfilled, but (B) not. Here we have $\kappa(C^{-1}K) = \mathcal{O}(h^{-2})$. So the number of iterations would grow with h^{-1} not optimally.

The modern preconditioning techniques such as

- the domain decomposition preconditioner (i.e. local preconditioners for interior degrees of freedom combined with Schur-complement preconditioners on the coupling boundaries, see [1,3,4,5,6,7])
- hierarchical preconditioners for 2D problems due to Yserentant [9]
- Bramble-Pasciak-Xu-preconditioners (and related ones see [1,8]) for hierarchical meshes in 2D and 3D

and others have this famous properties. Here (A) and (B) are given from the construction and from the analysis. The propety (C) is surprisingly fulfilled perfectly. Nearly the same is true, when Multigrid–methods are used as preconditioner within PPCGM, but from the inherent recursiv work on the coarser meshes, we cannot achieve exactly one data exchange over the coupling boundaries per step but L times for an L level grid.

Remark 3.5. All these modern preconditioners can be found as special cases of the Multiplicative or Additive Schwarz Method (MSM/ASM[8]) depending on various splittings of the f.e. subspace $\mathbb{V}_h$ into special chosen subspaces.

4. An Example for a Parallel Preconditioner

The most simple but efficient example of a preconditioner fulfiling (A), (B), (C) is YSERENTANT's hierachical one. Here we have generated the fine mesh from L levels subdivision of a given coarse mesh. One level means subdividing all triangles into 4 smaller ones of equal size (in the most simple case). Then, additionally to the nodal basis Φ of the fine grid, we can define the so called hierarchical basis $\Psi = (\psi_1, \cdots, \psi_N)$ spanning the same space $\mathbb{V}_h$. So there exists an $(N \times N)$-matrix Q transforming Φ into Ψ:

$$\Psi = \Phi Q.$$

From the fact that a stiffness matrix defined with the base Ψ

$$K_H = (a(\psi_j, \psi_i))_{i,j=1}^N$$

would be much better conditioned but nearly dense, we obtain from $K_H = Q^T K Q$ the matrix $C = (QQ^T)^{-1}$ as a good preconditioner:

$$\kappa(K_H) = \kappa((QQ^T)K) = \kappa(C^{-1}K).$$

From [9] the multiplying $\underline{w} := QQ^T \underline{r}$ can be very cheaply implemented if the level by level mesh generation is stored within a special list. Suprisingly, this multiply is perfectly parallel, if the lists from the mesh subdivision are stored locally in P_s (mesh in $\overline{\Omega}_s$).

Let $\underline{w} = Q\underline{y}$, $\underline{y} = Q^T \underline{r}$, then the multiply $\underline{y} = Q^T \underline{r}$ is nothing else than transforming the functional values of a "residual functional" $r_i = \langle r, \varphi_i \rangle$ with respect to the nodal base functions into functional values with respect to the hierarchical base functions: $y_i = \langle r, \psi_i \rangle$.

So this part of the preconditioner transforms type-II-data into type-II-data without communication and $\underline{y} = \sum H_s^T \underline{y}_s$.

Then the type-II-vector $\underline{y}$ is assembled into type-I

$$\tilde{\underline{y}} = \underline{y}, \quad \tilde{\underline{y}}_s = \underline{y}_s + \sum_{j \neq s} \underbrace{H_s H_j^T \underline{y}_j}_{\text{from other processors}}$$

containing now nodal data, but values belonging to the hierarchial base. So the function

$$w = \Phi\underline{w} = \Psi\tilde{\underline{y}}$$

is represented by $\underline{w}$ after back transforming

$$\underline{w} = Q\tilde{\underline{y}}$$

This is again a transformation of type-I-data into itself, so the preconditioner requires exactly one data exchange of values belonging to coupling boundary nodes per step of the PPCGM iteration.

Remark 4.1. For better covergence, a coarse mesh solver is introduced. Additionaly use of Jacobi-preconditioning is a worthy idea for beating jumping coefficiens and varying mesh spacings etc., so

$$C^{-1} = J^{-1/2} Q \begin{pmatrix} C_o^{-1} & \mathbb{0} \\ \mathbb{0} & I \end{pmatrix} Q^T J^{-1/2}$$

with $J = \text{diag}(K)$.

Remark 4.2. The better modern BPX-Preconditioners are implemented similarly fulfiling (A), (B), (C) in the same way.

5. Examples in Linear Elasticity

Let us demonstrate the power of this parallel finite element code at a 2D and 3D benchmark example.

We have used the GC Power-Plus-128 parallel computer at TU Chemnitz having up to 128 processors PC601 for computation and 4 T805 transputers for message passing per double-processor-node. The maximal arithmetical speed of 80 MFlops/s is never achieved for our typical application with unstructured meshes. Here, the matrix vector multiply $\underline{v}_s := K_s \underline{q}_s$ (locally at the same time) dominates the computational time. Note that $\vec{K}_s$ is a list of non-zero elements together with column index for efficient storing this sparse matrix.

We consider the elastic deformation of a dam. The following Fig. 1 shows the 1–level–mesh, so the coarse mesh contained 93 quadrilaterals.

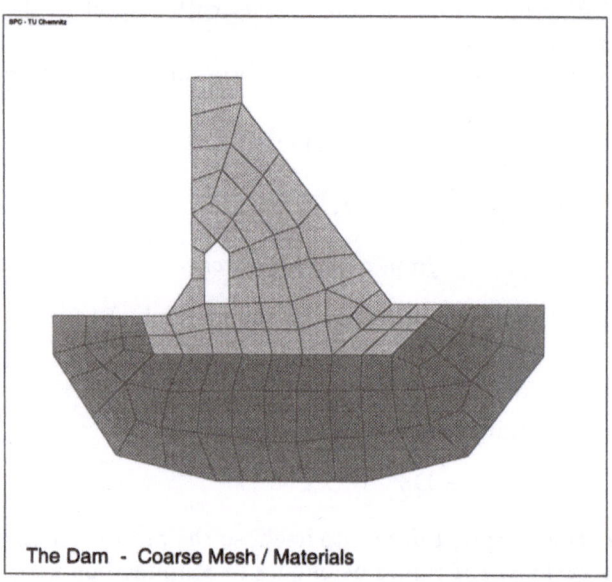

The Dam - Coarse Mesh / Materials

From distributing over $p = 2^m$ processors, we obtain subdomains with the following number of coarse quadrilaterals.

p	# quadrilaterals	max.speed–up
p=1	93	–
p=2	47	1.97
p=4	24	3.87
p=8	12	7.75
p=16	6	15.5
p=32	3	31
p=64	2	46.5
p=128	1	93

It is typical for *real life* problems that the mesh distribution cannot be optimal for a larger number of processors. In the following table we present some total running times and the measured percentage of the time for communication for finer subdivisions of the above mesh until 6 levels.

L	N	It	p=1	p=4	p=16	p=64
1	3,168	83	2.1"/0%	1.6"/75%	2.4"/90%	
2	12,288	100	9.2"/0%	3.7"/40%	3.3"/70%	
3	48,384	111	39.8"/0%	11.7"/20%	6.1"/50%	
4	192,000	123	–/–	46.2"/10%	16.9"/25%	
5	764,928	134	–/–	–/–	64.1"/10%	19.4"/25%
6	2,085,248	144	–/–	–/–	–/–	85.5"/9%

The most interesting question of scalabillity of the total algorithm is hard to analyze from this rare time measurements of this table. If we look at the two last diagonal entries, the quotient 85.5/64.1 is 1.3. From the table before and the growth of the iteration numbers a ratio of $(4 * 15.5/46.5) * (144/134) = 1.4$ would be expected, so we obtained a realistic scale–up of near one for this example.

The percentage of communication tends to zero for finer discretizations and constant number of processors. Much more problematic is the comparison of two non–equal processor numbers, such as 16 and 64 in the table. Certainly, the larger number of processors requires more communication start–up's in the dot–products ($log_2 p$). Within the subdomain communication the start–up's can be equal but need not. This depends on the resulting shapes of the subdomains within each processor, so the decrease from 10% to 9% in the above table is typical but very dependend on the example and the distribution process.

Whereas such 2D examples give scale–up values of near 90% for fine enough discretizations, much smaller values of the scale–up are achieved in 3D. The reason is the more complicate connection between the subdomains. Here, we have the crosspoints, the coupling faces belonging to 2 processors as in 2D but additionally coupling edges with an unstuctured rich relationship between the subdomains. So the data exchange within the preconditioning step (5) of PPCGM is much more expensive and 50% communication time is typical for our parallel computer.

A 3D benchmark of linear elasticity can be found at the web–site //www.tu-chemnitz.de/~tap/po3d.html (see:"Benchmark"). The domain Ω is the so called Fichera corner having 7 coarse cubes subdivided into 6 tetrahedrons each.

References

1. J. H. Bramble, J. E. Pasciak, A. H. Schatz (1986-89): The Construction of Preconditioners for Elliptic Problems by Substructuring I – IV,
   Mathematics of Computation,
   47,(1986) 103–134,
   49,(1987) 1–16,
   51,(1988) 415–430,
   53,(1989) 1–24.
2. I. H. Bramble, J. E. Pasciak, J. Xu (1990): Parallel Multilevel Preconditioners,
   Math. Comp. **55** 191, 1-22.
3. G. Haase, U. Langer, A. Meyer (1991): The Approximate Dirichlet Domain Decomposition Method,
   Part I: An Algebraic Approach,
   Part II: Application to 2nd-order Elliptic B.V.P.s,
   Computing **47** 137-151/153-167.
4. G. Haase, U. Langer, A. Meyer (1992): Domain Decomposition Preconditioners with Inexact Subdomain Solvers,
   J. Num. Lin. Alg. with Appl. **1** 27-42.
5. G. Haase, U. Langer, A. Meyer (1993): Parallelisierung und Vorkonditionierung des CG-Verfahrens durch Gebietszerlegung,
   in: G. Bader, et al, eds., *Numerische Algorithmen auf Transputer-Systemen*,
   Teubner Skripten zur Numerik,
   B. G. Teubner, Stuttgart 1993.
6. G. Haase, U. Langer, A. Meyer, S.V.Nepommnyaschikh (1994): Hierarchical Extension Operators and Local Multigrid Methods in Domain Decomposition Preconditioners,
   East-West J. Numer. Math. **2** 173-193.
7. A. Meyer (1990): A Parallel Preconditioned Conjugate Gradient Method Using Domain Decomposition and Inexact Solvers on Each Subdomain,
   Computing **45** 217-234.
8. P.Oswald (1994): Multilevel Finite Element Approximation: Theory and Applications,
   Teubner Skripten zur Numerik, B.G.Teubner Stuttgart 1994.
9. H. Yserentant (1990): Two Preconditioners Based on the Multilevel Splitting of Finite Element Spaces,
   Numer. Math. **58** 163-184.
10. L.Gross, C.Roll, W.Schönauer (1994): Nonlinear Finite Element Problems on Parallel Computers,
    In:Dongarra, Wasnievski, eds.,
    Parallel Scientific Computing, First Int. Workshop PARA '94, 247-261.

Employing Deterministic and Stochastic Petri Nets for the Analysis of Usage Parameter Control in ATM-Networks

Bruno Müller-Clostermann

Universität GH Essen, FB6/Informatik, D-45117 Essen
bmc@informatik.uni-essen.de

Abstract. Traffic flow control mechanisms play an important role for the design and operation of future high-speed networks. Here we employ the class of Deterministic and Stochastic Petri Nets (DSPN) for the specification and evaluation of Usage Parameter Control at the User Network Interface in ATM-networks. After an overview of performance issues in traffic management DSPNs are applied for the specification of the functional and quantitative behaviour of traffic sources and control mechanisms. The main contribution of this paper is to demonstrate the applicability of DSPN as a concise and unifying technique for the specification and analysis of Usage Parameter Control in ATM-networks.

1 Introduction

Different activities that are vital during the development process of complex distributed systems may be supported by formal methods and tools. Formal specification techniques are increasingly applied in the area of telecommunication and distributed systems. However, the use of quantitative evaluation techniques to asses the performance behaviour of a system under design is still an exception. The major reason is that performance evaluation has been (and is still) isolated from the methodology of system design. In his famous article [8] discussing the insularity of performance evaluation Domenico Ferrari even uses the term esoteric cult. Moreover, Ferrari proposes to foster the integration of the performance evaluation viewpoint with the mainstream of computer and communication systems engineering.

Today, the need for the integration of functional and quantitative techniques is undisputed. Especially the use of a formal system specification as starting point for quantitative evaluations is widely accepted as a promising approach to consider performance aspects during the design phase. In particular in the area of interactive systems, real-time and embedded systems, and in telecommunication systems performance aspects are closely related to functional correctness [11].

We contribute to this newly developing methodology of integrated quantitative analysis using Deterministic and Stochastic Petri Nets (DSPN) as a concise and unifying technique for the specification and performance analysis of certain flow control problems in high speed networks.

Here we consider the so called *leaky bucket* flow control that plays an important role in traffic management of ATM-networks. We employ timed Petri nets as a unifying method for the formal specification and performance analysis of traffic sources and *usage parameter control (UPC)* at the user network interface (UNI) of ATM networks. Functional as well as performance aspects are described by *Deterministic and Stochastic Petri Nets (DSPN)*. In particular quantitatively assessable models are obtained that may be investigated by numerical and simulative solution techniques. Presentations of some experimental results using the Petri net analysis tools DSPNexpress [23] and TimeNET [10] conclude the paper.

2 Performance Issues in ATM-networks

The *Asynchronous Transfer Mode (ATM)* is the transport mode for the future Broadband-ISDN (B-ISDN). ATM is capable of multiplexing a large number of connections providing services like voice, data, TV, video and others. Due to complex traffic characteristics in ATM-networks a wide range of problems must be resolved before such networks can be effectively managed. These problems include many performance issues that have been attacked by advanced mathematical techniques [2], [4], [19], [30], [33], [34], [38], [12]. One of the main topics in ATM networks is *traffic management* that has to provide for effective congestion control to prevent the network becoming a bottleneck. The congestion control techniques proposed for ATM cover a wide range of time scales, including the (re)definition of virtual paths, traffic shaping by buffering cells to reduce the peak rate, call admission and bandwidth allocation, and usage parameter control or traffic policing. Of outstanding importance in this context is the characterization of traffic sources being a permanent major research subject in the telecommunication area. An overview on recent developments in the area of ATM traffic management is given in text books [5], [32], [35] and numerous conference and journal papers, see e.g. [4], [31], [20], [12].

In the next sections we shortly review some issues concerning traffic characterization, *call admission control (CAC)* and especially *usage parameter control (UPC)*. Note that we use the terms UPC and policing function as synonyms.

2.1 Traffic Sources

The ATM-concept comprises different traffic sources like voice, data, video conferencing and TV within a multi-service environment. The channel is a resource shared among many users according to a time multiplexing scheme.

The bit streams generated by the sources are divided into cells which have a length of 48 byte supplemented by a 5 byte header. The ATM-header fields carry information concerning flow control, load type, priority, error control, and identifiers for virtual paths and virtual channels. Note that a virtual path (VP) is a collection of virtual circuits (VCs) between two nodes. Hence the length of the data part of a cell in bits is $n_{cell} = 48 \cdot 8 = 384$ and a bit rate of e.g.

1M [bits/sec] yields a net cell rate of $2^{20}/384 = 2730.6$ [cells/sec]. In Gigabit networks traffic rates of millions of cells per second may occur.

The characterization and modeling of sources is one of the central topics in high speed network performance. For an overview see e.g. [5], [30], [32]. A widely used set of traffic parameters is

B_p	peak bit rate [bits/sec]
	eqv. to peak cell rate $C_p = B_p/n_{cell}$ [cells/sec]
B_m	mean bit rate [bits/sec]
	eqv. to mean cell rate $C_m = B_m/n_{cell}$ [cells/sec]
$b = B_p/B_m$	burstiness of the traffic source
T	mean burst duration [sec]
	eqv. to mean burst length $L = T/(n_{cell}/B_p)$ [cells]
$T \cdot (b-1)$	mean silence duration

The *burstiness* b of a source is defined as the peak to mean bit rate ratio, i.e. $b = B_p/B_m$. By definition a constant bit rate source (CBR source) like uncompressed video or voice where silence periods are transmitted has a *burst factor* of $b = 1$. Variable bit rate sources (VBR sources) like bulk file transfer, compressed video and voice with silence detection have burst factors $b > 1$.

Note that the burst duration together with the peak bit rate defines the length of a burst measured in cells. The average burst length L relates to the other parameters through the equation $L = T/(n_{cell}/B_p)$.

Typically so called on/off-models are used for the characterization of bursty sources. Also the more complex multi-state models are used. We will take up the issue of traffic source modeling in a later section.

When studying the problem of call acceptance control usually a mix of bursty sources is considered that share a transmission link with limited buffer size. A typical problem is to find the link bandwidth W that has to be assigned to this traffic mix in order to achieve a required *quality of service (QoS)*. The major QoS-parameter is the cell loss probability P that is required to range from 10^{-4} to 10^{-10} depending on the type of service. A problem related to call acceptance control is the usage parameter control or policing discussed in the next section.

2.2   Usage Parameter Control

During the call admission procedure the source has to supply a set of parameters that become a part of the *traffic contract* between the user and the network. Based on this contract the network will try to control the sources in order to minimize the potentially negative effect of the excessive traffic of non conforming sources to the conforming ones. Hence nonconforming sources should be detected as quickly as possible and appropriate actions like dropping, marking or extra charging of violating cells have to be taken. This issue in ATM-systems is refered to as usage parameter control or source policing. It is located at the *user network interface (UNI)* that forms the access point to the network [32], [35]. The same mechanisms may be used at the network network interface (NNI).

There may be several reasons that sources do not stay within the negotiated connection parameters. The user equipment may show malfunctions or the user may even try to provocate a network crush. The major reason however is that users underestimate their requirements concerning mean or peak bandwidth. As a consequence the sources may not stay within the connection parameters negotiated at the call-set-up-phase.

In the scheme presented in fig. 1 cells passing the UPC procedure are multipexed over the same channel that provides a specified bandwidth for a set of sources. Note that sources may either be policed individually or in groups. The latter is the case if a carrier who leases VPs (virtual paths) for private networks is not interested in policing individual VCs (virtual channels) multiplexed onto these VPs [32], p. 137.

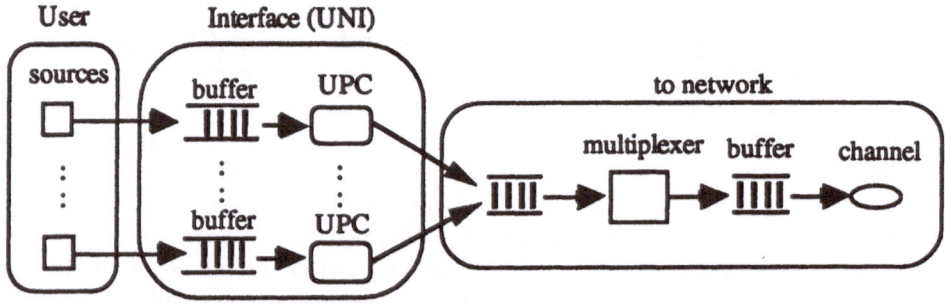

Fig. 1. Usage Parameter Control in ATM-networks

We shortly discuss some issues of UPC. For details we refer to [32], [34], [35].

Apart from the mean bit rate, also the peak bit rate and the length of a burst are parameters that may be controled by UPC. Since a policing function can control only one parameter, dual or even multiple schemes have been proposed, either in serial or in parallel configuration. In case of a serial scheme the first function may police the peak rate, the second the average rate.

The peakiness of the traffic characterized by the burst factor is obviously of outstanding importance. Usually a policing function should be *transparent* to the source, i.e. cells should not suffer any additional delay due to the policing function. Depending on the duration or distributions of the burst periods a buffered scheme with *traffic shaping* properties may be preferable. In this case an additional delay in the cell buffer must be tolerated. An example is the *buffered leaky bucket (BLB)* that may be appropriate for highly bursty sources.

The bit rates negotiated during call admission are average measures defined over a certain interval of time. E.g. the true-average cell rate denotes the number of cells generated during the connection duration divided by the length of the duration [32], p. 137. Since this value is not known in advance the policing of the true-average cell rate is not possible and a (virtual) average cell rate defined over an appropriate interval of time must be policed. In some cases it may be reasonable to police a source at different load levels and/or timescales, e.g. a scheme of multiple leaky buckets with different bucket sizes may be employed [13].

The main measure of interest in the quantitative evaluation of UPC schemes is the *violation probability* P_v, i.e. the probability that a cell from a conforming source will be discarded. Another measure of interest could be the amount wasted bandwidth that has been reserved but has remained unused. The reaction time for an UPC to detect a nonconforming situation may be another evaluation criterium. Apart from the violation probability P_v, also other measures for erroneous reactions of the UPC may be of interest.

2.3 UPC Schemes

The technique recommended by the ITU is the leaky bucket that exists in several versions. The basic idea is that an arriving cell is forwarded to the network only if at least one token (or credit) is available in a token pool. In this case the token is consumed. Tokens are generated at a constant rate and the number of tokens in the token pool must not exceed an upper limit, otherwise tokens are lost. If an arriving cell does not find a token it is discarded (or tagged) or in case of a buffered leaky bucket it is stored in a cell buffer. The leaky bucket scheme exists in several variants that are usually described in the literature by informal or semi-formal notions. Fig. 2 shows an informal description of two leaky bucket schemes.

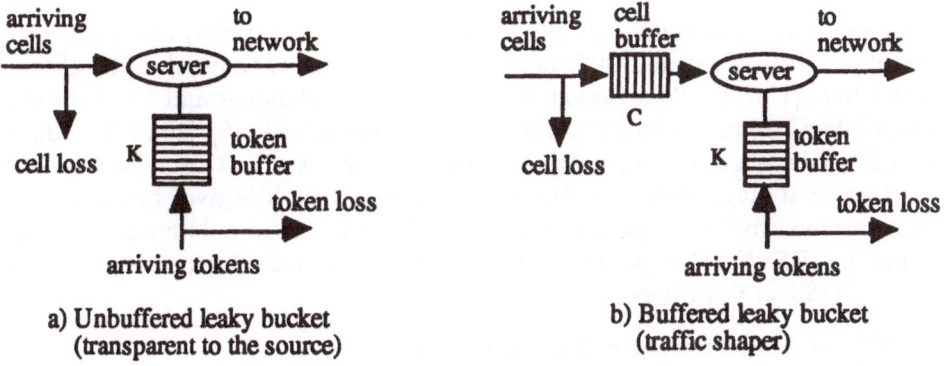

a) Unbuffered leaky bucket
(transparent to the source)

b) Buffered leaky bucket
(traffic shaper)

Fig. 2. Leaky Bucket Schemes

A leaky bucket is specified by three parameters: token generation rate R (also called leaking rate), size K of the token pool, and in case of a buffered scheme, the size C of the cell buffer. One of the major objectives in leaky bucket design is the dimensioning of the *bucket size* and the amount of additional bandwidth represented by an *overdimensioning factor*. In the unbuffered case the token pool size K and the overdimensioning factor f have to be determined such that the violation probability P_v is smaller than a predefined tolerance, e.g. $P_v \leq 10^{-9}$. In the buffered case additionally the trade off between the cell buffer size C and the token pool size K must be considered.

Window based techniques observe the number of cells arriving during a certain time interval. If the limit is reached further cells are discarded. The literature reports on *Jumping Window*, *Triggered Jumping Window* (also called Stepping Window) and *Moving Window*. The two parameters of window based schemes are the window size and the upper limit of the cell counter.

3 Deterministic and Stochastic Petri Nets

Petri nets have been traditionally used for the design and functional analysis of concurrent and parallel systems. Properties like boundedness, deadlock or liveness may be investigated on the basis of a Petri Net description. Also the interactive validation by functional simulation, called token game is of great value.

Since pure Petri Nets do not know the concept of time quantitative analyses concerning performance and reliability issues are not possible. During the last decade a number of proposals for the inclusion of time in Petri nets have been made leading to *Deterministic and Stochastic Petri Nets (DSPN)*. Nowadays there are abundant techniques and tools available decribing these type of models and their application [1], [26], [28].

3.1 An overview of DSPN

Here we survey Deterministic and Stochastic Petri Nets (DSPN) that have been introduced by Marsan and Ciola in 1987 as an extension of *Generalized Stochastic Petri Nets (GSPNs)* [28]. Recent contributions to techniques and toosl are due to Ciardo, German, Kelling and Lindemann and others [3], [9], [10], [17], [18]. A DSPN-model is a directed bipartite graph. The first set of vertices P corresponds to places that may hold tokens; the second set of vertices T is given by transitions that may be timeless or associated either with a random or a deterministic time delay, i.e. DSPN allow for three different types of transitions given by a set $T = T^Z \cup T^E \cup T^D$, where

- T^Z are immediate (=timeless) transitions
- T^E are transitions with an exponentially distributed firing time. Hence, each $T_i \in T^E$ is associated with a firing rate λ_i.
- T^D are deterministic transitions with constant firing times. Hence, each $T_i \in T^D$ is associated with a constant time duration d_i.

There are input, output and inhibitor arcs. With each arc is associated its multiplicity. Input arcs and inhibitor arcs connect places to transitions and output arcs connect transitions to places. A transition is enabled if all its input places contain at least as many token as the multiplicity of the corresponding input arc and all of its inhibitor places contain fewer token as the multiplicity of the corresponding inhibitor arc. An enabled transition fires by removing tokens from the input places and adding tokens to the output places according to the arc multiplicities. The DSPN-constructs displayed in fig. 3 show some of the graphical symbols used in DSPN.

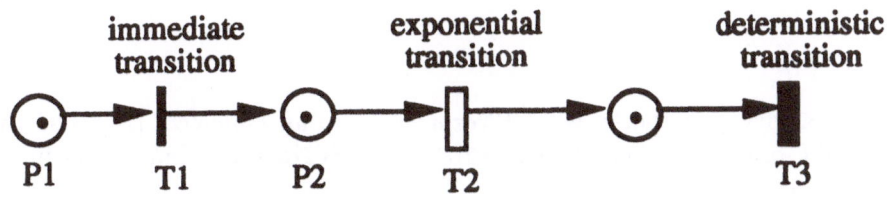

Fig. 3. Graphical Representation of DSPN-Elements

In fig. 3 the transitions T_1, T_2, and T_3 are concurrently enabled. As known from GSPN immediate transitions have priority over timed transitions. Hence, transition T_1 fires immediately and the actual marking (1,1,1) changes immediately to (0,2,1). The first marking is called vanishing, the new one is a tangible marking. Afterwards T_2 or T_3 may fire; in case the exponential transition T_2 wins the race against the deterministic transition T_3, the marking (0,1,2) is reached, otherwise T_3 will fire resulting in the marking (0,2,0). Eventually both tokens are swallowed by T_3.

To resolve firing conflicts between immediate transitions priorities may be specified for all $t \in T^Z$. Fig. 4 shows a loss model of type M/M/1/3, i.e. inter-arrivals from the source and service durations are exponentially distributed and the capacity is restricted to 3.

Due to the priority of T_3 over T_2 losses can occur only if T_3 is not enabled, i.e. if P_3 contains exactly three tokens and P_2 is empty.

Another concept of DSPN to influence conflicting timeless transitions are weights w_i for $t_i \in T^Z$. Weights are used to determine the firing probabilities under a given marking. Moreover, weights, firing delays, and arc multiplicities may be marking dependent.

3.2 Evaluation Techniques for DSPN-Models

DSPN-models may be analyzed either functionally by the classical Petri-net techniques like token game, place- and transition invariants, and reachabilty analysis, or quantitatively using numerical techniques or discrete event simulation. Here

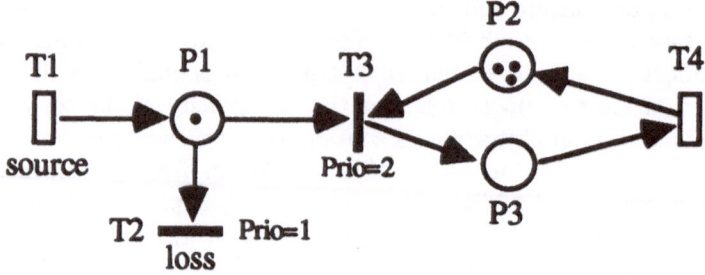

Fig. 4. Controlling Conflicts using Priorities

we give a survey of the tools *DSPNexpress* [23] and *TimeNET* [16]. The tool TimeNET originates from the same roots as DSPNexpress, but differs in the provided techniques.

Both tools provide a comfortable graphical user interface and numerical solution techniques for stationary and transient analyses based on the Markov chain underlying the DSPN-model. Interactive token game and functional analyses for the determination of invariants and conflicting transitions serve for validation purposes. For quantitative evaluation the *steady state probability distribution* of the reachability set is computed and measures like throughput and token probability distribution are determined.

Both tools provide advanced numerical solution techniques. In case of pure bounded SPNs (i.e. we have a finite reachability set and all transitions are exponential) the model is isomorphic to a finite continuous time Markov chain (CTMC) that can be solved via the global balance equations. If both immediate and exponential transitions are employed we have a model of the GSPN-type showing tangible and vanishing states. Before computing the steady state distribution the vanishing states are eliminated [27], [15]. The same technique holds for models with multiple arcs, inhibitor arcs, priorities and transitions weights. If a model contains deterministic transitions the memoryless property does not hold any longer. In this case the technique of subordinated Markov chains is used that evaluate states where exponential transitions are in conflict with the deterministic one, for details see [3], [28]. The evaluated measures are the throughputs of timed transitions and the probabilities of the place populations. Numerical techniques for *transient analyses* are also available.

Sometimes a numerical solution is not feasible, e.g. if there are conflicting deterministic transitions or a very large (or even infinite) state space. Discrete event simulation as supplied by TimeNET is an alternative. Fast simulation techniques based on importance sampling for the analysis of models with rare events [18] as well as distributed simulation in workstation clusters [17] are part of TimeNET's evaluation techniques.

4 Specification of Traffic Sources with SPN

Here we introduce the class of *Markovian Arrival Processes (MAP)* that are very convenient for the stochastic modelling of traffic sources in telecommunication networks. MAPs form a hierarchy that includes simple Poisson sources as well as arbitrarily detailed multi-state models. We present some example specifications, where all transitions are exponential, i.e. here the subclass of *Stochastic Petri nets (SPN)* will be used for traffic modelling.

4.1 Markovian Arrival Processes and its Subclasses

The versatility of the very general class of Markovian Arrival Processes (MAPs) for the modelling of traffic sources in future telecommunication systems has been shown in many studies. MAPs include as subclasses the Markov *Modulated Poisson Processes (MMPP)* and the *Interrupted Poisson Processes (IPP)*. An IPP (Interrupted Poisson Process) switches between the two states active and passive. In its active state arrival events are generated with rate $\lambda > 0$, otherwise it is silent (i.e. $\lambda = 0$). An IPP is often used for modelling packetized voice and other burst-silence sources.

Also phase-type distributions, Cox-distributions, and hence the hyper-exponential and the Erlang distributions are covered by the MAP-definition. At the bottom of the hierarchy we see the Poisson process, denoted by the letter M which stands for Markovian. Note that the classes MAP and MMPP are closed with respect to independent superposition, e.g. merging several independent MMPPs leads again to an MMPP.

These processes are doubly stochastic processes whose arrival rates are functions of the state of a finite irreducible Markov chain. For definitions and properties of IPP, MMPP and MAP see also [25], [32]. Here we present definitions of IPPs and MMPPs in terms of SPNs.

4.2 Modelling IPPs and MMPPs as SPNs

Since Stochastic Petri nets (SPNs) provide only exponential transitions they form a subclass of DSPNs. SPNs are known to be isomorphic to time-continuous Markov chains [27], in particular Poisson processes, IPPs and MMPPs can be straightforward represented as SPN-models. Fig. 5 displays three examples.

As shown in fig. 5a) the Poisson source is a one-state source. Note that the cycle formed by place S_0 and its arc may be omitted. A k-state MMPP is a doubly stochastic process that captures both the time varying arrival rates and the correlation between the interarrival times, cf. fig. 5c). The places P_0, P_1, ..., P_k represent the feasible states of the source that are changed according to rates $\alpha_i, i = 1, 2, ..., 2k$. Exactly one token circulates between the places P_i indicating the actual state of the source. In state S_i the source is active with a send rate of λ_i. The place P' may serve as an interface to connect the source to a subnet.

For the modelling of voice sources with silence detection, usually an IPP also called on/off-process is used. We assume that both active and passive periods

110

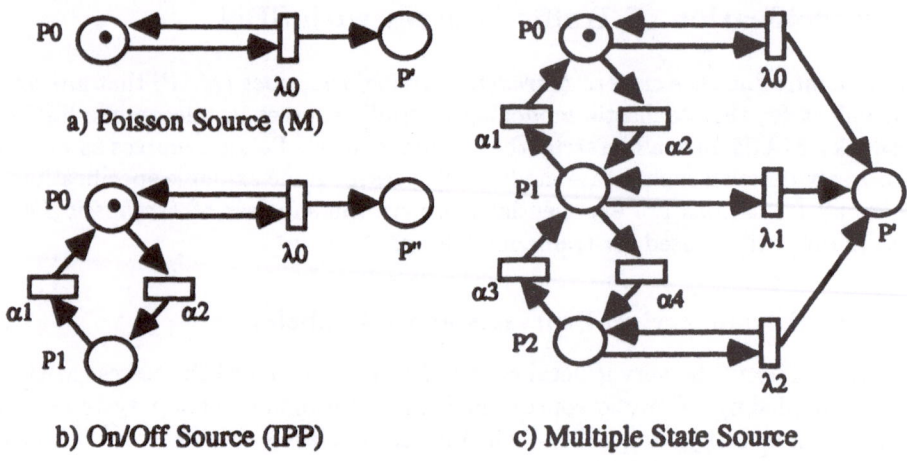

a) Poisson Source (M)

b) On/Off Source (IPP) c) Multiple State Source

Fig. 5. Stochastic Petri Nets modelling Traffic Sources

are exponentially distributed with average time durations T and $T \cdot (b-1)$ respectively. Hence, the IPP-parameters are obtained immediately by the contract parameters B_p, B_m, and T, cf. section 2.1. Note that the source models use cell rates C_p and C_m instead of bit rates B_p and B_m.

The traffic source parameters C_p and C_m, T are directly related to the IPP-parameters by the formulas given in notion 1.

peak cell rate	$C_p = \lambda_0$
mean cell rate	$C_m = \lambda_0 \cdot \text{Prob}[S_0] = \lambda_0 \cdot \alpha_1/(\alpha_1 + \alpha_2)$
burst factor	$b = (\alpha_1 + \alpha_2)/\alpha_1$
mean burst duration	$E[T] = 1/\alpha_2$
mean burst length	$E[L] = \lambda_0 \cdot E[T] = \lambda_0/\alpha_2$

Notions 1. Source Parameters and their Relation to IPP-Parameters

For more complex traffic patterns multi-state sources of the MMPP type are appropriate. An example for this type of traffic is compressed video that has a variable bitrate (VBR). In case of audio/video or multimedia MAPs or MMPPs may not adequately represent the traffic sources. As a consequence more details may be included and other evaluation approaches should be used, even the usage of traces obtained from real video sequences are usable in simulations [37].

5 Modelling of Usage Parameter Control by DSPN

Here we describe some UPC schemes by means of Deterministic and Stochastic Petri Nets. After leaky bucket schemes, their evaluation, and possible extensions some window techniques are briefly sketched. The objective of the investigation is the evaluation of stationary performance measures, in particular the *cell loss probability* that in the given context is denoted as violation probability.

5.1 Leaky Bucket Schemes

A token buffer is filled with credits at a constant rate. An ATM-cell is passed to the network if a credit (=token) is available, otherwise it is lost. Credits may get lost too; in this case some of the reserved bandwidth has not been used by the source.

For the buffered case the bucket size is given by $C + K$, cf. fig. 2b). A large value of C accounts for additional delays and traffic shaping effects whereas in case of a small C the occuring bursts are delivered (almost) transparently to the network, cf. [32], p. 141.

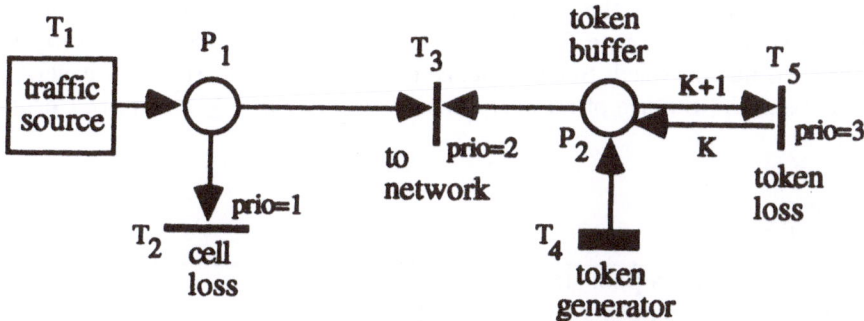

Fig. 6. DSPN-model of the Unbuffered Leaky Bucket

Here we present a timed Petri Net description of a model comprising a traffic source and an UPC-mechanism that is formal, complete, unambiguous and assessable by functional and quantitative evaluation techniques. Due to the unified description of the traffic source and the policing scheme a qualitatively and quantitatively assessable model is obtained. In the following figures the traffic source is specified as a functional block that represents a Petri-Net submodel with exactly one output arc serving as input to the rest of the model. Note that all graphical descriptions have equivalent textual representations that are easily derivable.

Note that the use of priorities allows to control the conflicts due to multiple transition activations like in the case of transitions T_2 and T_3. In case of a non-empty token buffer (place P_2) a cell arrival will not lead to a loss because T_3 has priority over T_2. The notions are explained in detail in notion 2.

P_1	Holds at most one cell
P_2	Token pool of capacity K (represents the bucket size)
T_1	Submodel generating cells (e.g. on/off-source or multi-state-source)
T_2	Removes cells to be dropped (T_2 has priority 1, i.e. lowest priority)
T_3	Represents the interface to the network (T_3 has priority over T_2)
T_4	Token generator with deterministic timing R (R specifies the leaking rate)
T_5	Represents the loss of tokens (unused bandwidth)
prio	Priority of transitions

Notions 2. Leaky Bucket Parameters used in fig. 6

The parameters of the unbuffered leaky bucket are the token pool size K and the token generation rate R that specifies an upper limit for the cell rate (bandwidth) to be controled. Since a choice of $R = \lambda_m$ would result in a very large bucket size K, the use of an *overdimensioning factor* $f \geq 1$ has been proposed [34].

$K > 0$	token pool size
$R > 0$	token generation rate (=leaking rate)
$f \geq 1$	overdimensioning factor

Notions 3. Parameters of the Unbuffered Leaky Bucket

The state of the model, in terms of Petri nets the marking, is described by (s, m_1, m_2), where $s, s = 1, 2, \ldots, k$, denotes the state of the source and (m_1, m_2) denotes the state of the leaky bucket. For the purpose of performance evaluation only the tangling (=non-vanishing) markings are of interest. In case of a stateless Poisson source, the state space is given by the tangling markings

$(0, 0, 0), (0, 0, 1), \ldots, (0, 0, K)$ and a stationary analysis of this scheme will yield $K + 1$ steady state probabilities $\text{Prob}[m_2 = i], i = 0, 1, \ldots, K$.

For an on/off-source, the steady state probabilities are given as $\text{Prob}[s = i, m_2 = j] = \text{Prob}[s = i] \cdot \text{Prob}[m_2 = j]$. The main measure is the violation probability $P_v = \text{Prob}[s = 0] \cdot \text{Prob}[m_2 = 0]$. Measures of minor interest are the cell loss rate $C_l = \lambda_0 \cdot P_v$, the cell throughput $T_c = \lambda_0 \cdot (1 - P_v)$ and the token loss rate $T_l = R - T_c$ (where R denotes the token generation rate). The measures for multi-state-sources are expressible in a very similar way.

Fig. 7 shows the buffered leaky bucket scheme, where a cell buffer, represented in the DSPN-model by the place P_1, may hold up to C cells. Hence the set of tangible markings includes additionally the markings where $0 \leq m_1 \leq C$.

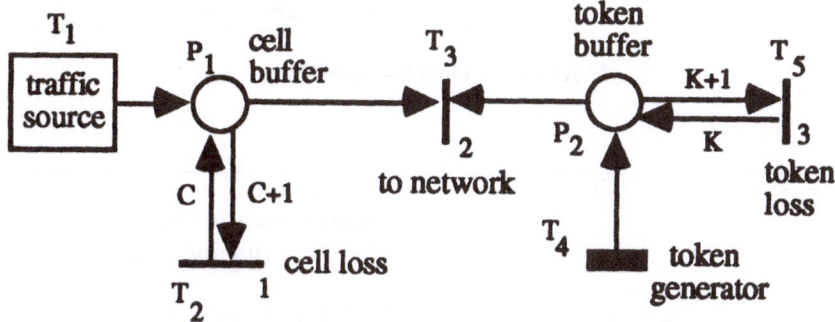

Fig. 7. DSPN-model of the Buffered Leaky Bucket

$K > 0$	token pool size
$C \geq 0$	cell buffer size
$R > 0$	token generation rate (=leaking rate)
$f \geq 1$	overdimensioning factor

Notions 4. Parameters of the Buffered Leaky Bucket (BLB)

An arriving cell will be dropped if P_1 holds its maximum number of C cells. The parameters of the buffered leaky bucket are C, K and R. Here the violation probability is given as $P_v = \text{Prob}[m_0 = \text{active}] \cdot \text{Prob}[m_1 = C, m_2 = 0] = \text{Prob}[m_0 = \text{active}] \cdot \text{Prob}[m_1 = C]$.

Other measures are the cell loss rate $C_l = \lambda_0 \cdot P_v$, cell throughput $T_c = \lambda_0 \cdot (1 - P_v)$, token loss rate $T_l = R - T_c$, and the mean cell delay $D = \text{E}[m_1] \cdot T_c$, where $\text{E}[m_1]$ ist the mean number of cells in the cell buffer and T_c is the cell throughput. Note that the non-transparency of the scheme increases with the delay D. The violation probability is a function of the bucket size $K + C$ and the leaking rate R, i.e. $P_v = f(K + C, R)$. Note that the buffered leaky bucket shows the same values P_v like the unbuffered case.

As an example for numerical evaluations we display in fig. 8 and 9 some results obtained for the buffered leaky bucket using the numerical parameter values shown in notions 5 and 6. In particular we show the effects obtained by increasing the overdimensioning factor from 1.0 to the value 1.2.

b	2	burst factor
C_p	20.0	peak cell rate
C_m	10.0	mean cell rate
T	0.2, 0.4, 0.8	mean burst duration

Notions 5. On/Off-Surce Parameters

$K + C$	10 ... 200	bucket size
R		token generation rate
f	1.0, 1.2	overdimensioning factor

Notions 6. Parameters of the Buffered Leaky Bucket

Obviously the violation probability P_v cannot be decreased to the desired values without overdimensioning of the allocated bandwidth. Note that the relation of K and C may be adapted according to the tolerable delay for the given service without any effect on the value of P_v.

5.2 Multiple Leaky Buckets

For the control of more than one measure multiple leaky buckets can be used. According to the negotiated contract parameters we have to parameterize the buckets by appropriate values for K, C, and f. In case the *peak bandwidth* defined over a rather short time interval as well as the *mean bandwidth* defined over a larger interval are both to be controled, a scheme of two *serial leaky buckets* may be used. A DSPN-model describing the situation is diplayed in fig. 10.

DSPNexpress as well as TimeNET provide evaluation techniques for the tuning of the bucket parameters under different traffic scenarios. The model assumes bucket sizes of 16 and 128 respectively and a two-state source yielding a rather modest model size of 4386 tangling states. Of course we have to fight the state space explosion problem in case of larger bucket sizes.

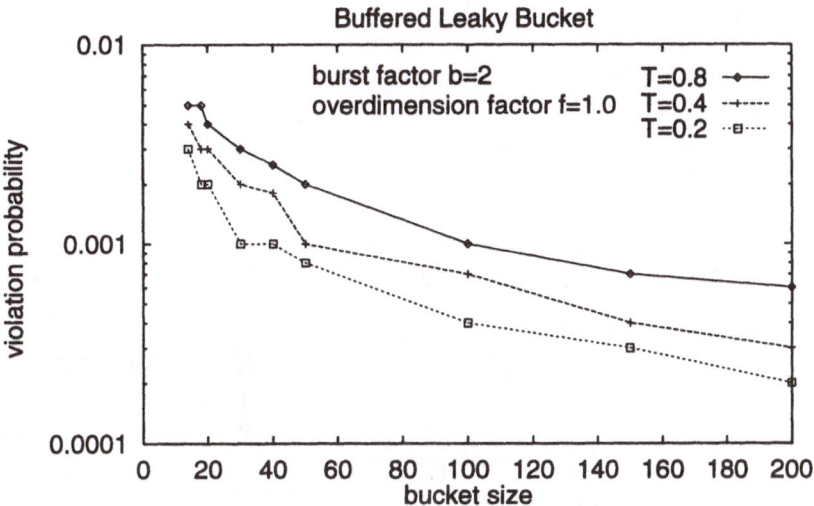

Fig. 8. Violation Probability of the BLB without Overdimensioning

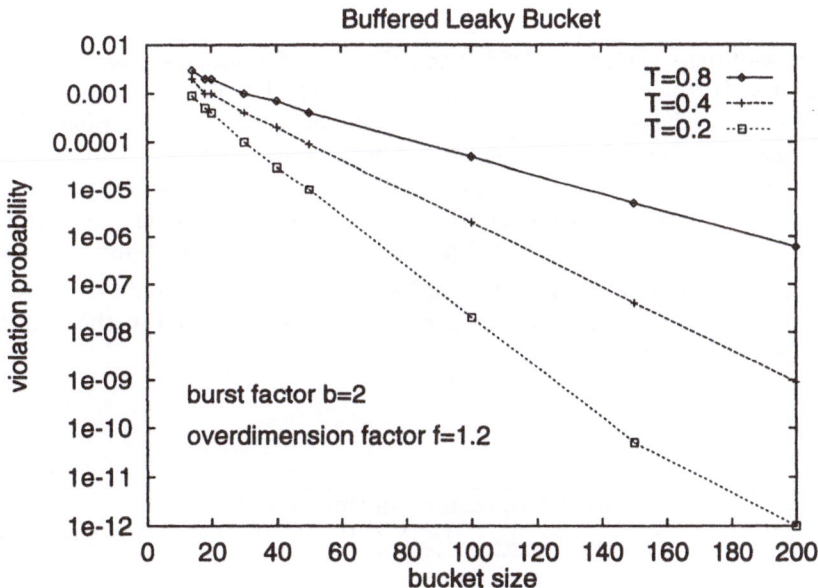

Fig. 9. Violation Probability of the BLB with Overdimensioning Factor f=1.2

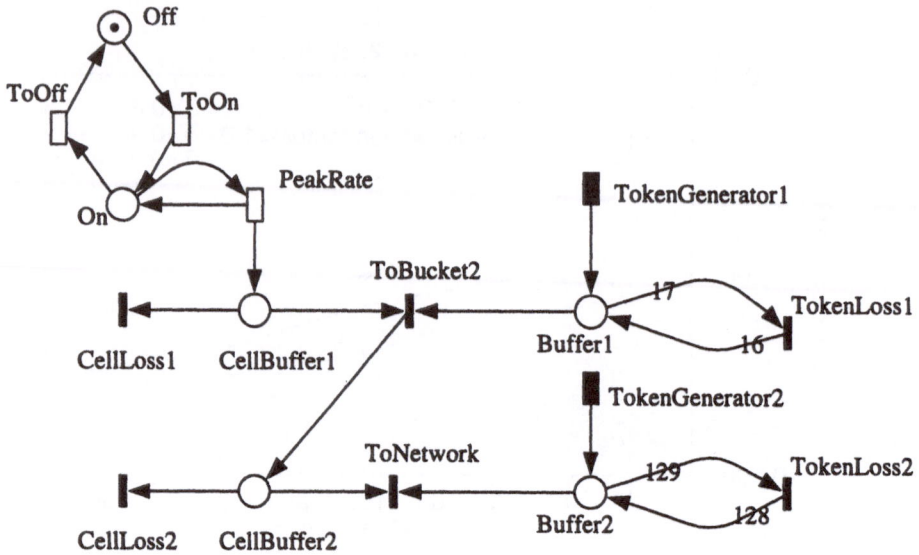

Fig. 10. DSPN-model of Serial Leaky Buckets

5.3 Window Based Techniques

Here we briefly sketch the DSPN-specifications for window based UPC. We provide the UPC-parameters and the P_v-formula. For detailed discussions of the window based scheme see [32]. The *Jumping Window technique (JW)* uses windows of fixed length. When a window expires the next window starts immediately, i.e. the window is not synchronized with the activity of the source. A counter initially set to 0 counts the number of arriving cells until the window expires. If the counter reaches a predefined limit newly arriving cells are lost.

The parameters of the Jumping Window are the window size W (measured in time units) and the number of cells maximally allowed in the window, here denoted by K, cf. notion 7.

$W > 0$	window size (measured in time units)
$K > 0$	cell counter limit, $0 \leq k \leq K$, (K/W is the controlled rate)

Notions 7. Parameters of the Jumping Window

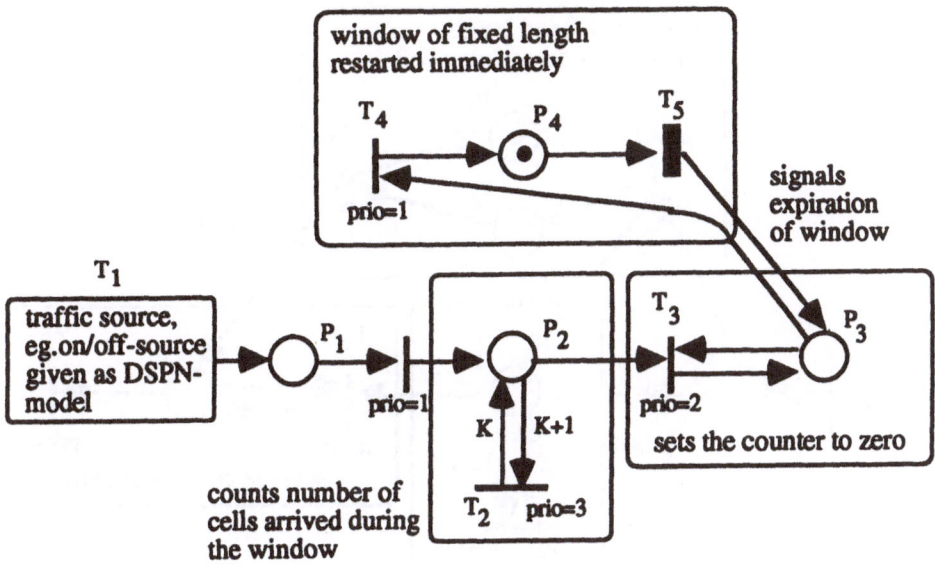

Fig. 11. DSPN-Model of the Jumping Window

Hence the mean cell rate controlled by this scheme is given bei $l = K/W$. Given a rate l and an arbitrarily chosen K, the window size is determined simply by $W = K/l$. As pointed out in [32], p.142, if W is increased it takes longer to detect that the controlled rate has been exceeded.

As W is decreased, cells may be dropped although the source does not violate the contract. In case of a multi-state-source the violation probability is given as $P_v = \mathrm{Prob}[\text{source is active}] \cdot \mathrm{Prob}[m_2 = K]$. In case of a Poisson source the first term equals 1.0.

In case of the *Triggered Jumping Window (TJW)* a new window starts with the next cell arrival. Hence this version is sometimes also called stepping window. Fig. 12 displays a DSPN-model of TJW. The parameters are window size W and cell counter limit K. The violation probability of this scheme and the other performance measures are computed like in case of the Jumping Window.

The *Moving Window* is known to be equivalent to a multiple server loss system of type M/D/K [34] that may also be described as a DSPN-model. Due to the multiple deterministic transitions a straightforward solution with numerical techniques is not straighforward. Analytical solutions exist for the simple baseline model.

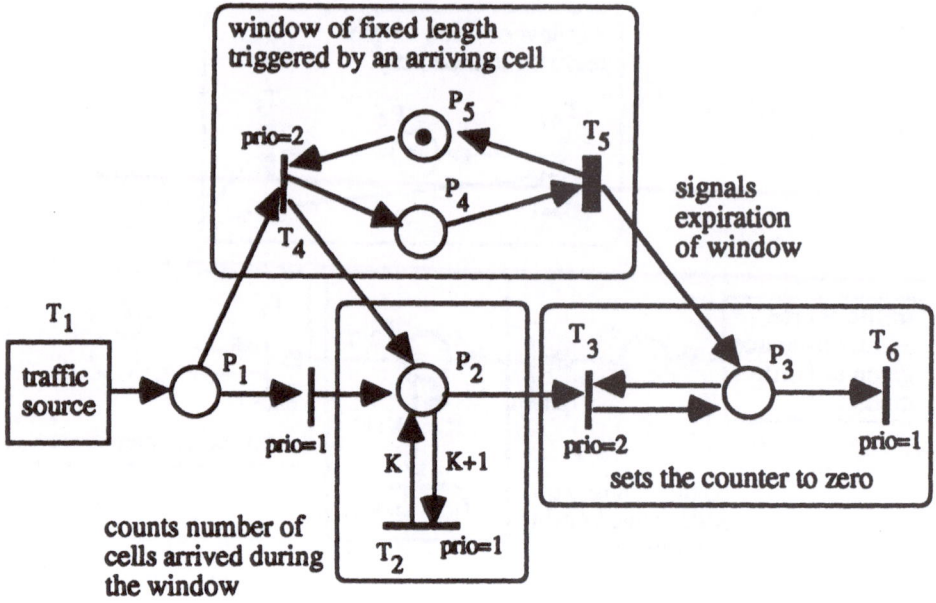

Fig. 12. DSPN-model of the Triggered Jumping Window

6 Conclusion and Outlook

We have shown how Deterministic and Stochastic Petri Nets may be used for the investigation of UPC-mechanisms. In particular a framework for the definition and investigation of leaky buckets has been developped.

We do not claim any advancements in technical or algorithmic details, but we hope to have contributed to methods for the management of telecommunication and networking systems.

An important area for future work is the investigation of transient UPC-behaviour. Especially stress situations caused by traffic bursts or superposition of a new connection and existing traffic will be of interest. Due to the large amount of cells that have to be moved under strict QoS requirements in short time intervals such investigations are a great challenge. Additional to numerical approaches also advanced simulation techniques including distributed and parallel simulation as well as measurements will be necessary.

7   Acknowledgements

The author thanks Frank Broemel, Kristine Karch, and Andreas Mindt for their technical support. The critic of the unknown referee helped to improve the paper.

References

1. Bause, F., Kritzinger, P.: Stochastic Petri Nets - An Introduction to the Theory. Advanced Studies in Computer Science, Vieweg (1995)
2. Butto, M., Cavallero, E., Tonnietti, A.: Effectiveness of the Leaky Bucket Policing Machanism in ATM Networks. IEEE Journal on Selected Areas in Communications, Vol. 9., No. 3 (April 1991)
3. Ciardo, G., German, R., Lindemann, C.: A Characterization of the Stochastic Process Underlying a Stochastic Petri Net. IEEE Trans. on Software Engineering, Vol. 20, No. 7 (July 1994) 506–515
4. Cohen, J.D., Pack, C.D. (eds.): Queueing, Performance and Control in ATM. Proc. of the ITC-13, North Holland (1991)
5. de Prycker, M.: Asynchronous Transfer Mode - Solution for Broadband ISDN. Ellis Horwood (1990)
6. Devetsikiotis, M., Townsend, J.K.: Statistical Optimisation of Dynamic Importance Sampling Parameters for Efficient Simulation of Communication Networks. IEEE/ACM Transactions on Networking, Vol. 1, Number 3 (June 1993) 293–305
7. Diavolitsis, H.G., Helvik, B.E., Lehnert, R., Michelsen, J.C.: Traffic Generation for ATM Systems Testing Environment Modelling and Feasability Studies. [4]
8. Ferrari, D.: Considerations on the Insularity of Performance Evaluation. IEEE Trans. on Softw. Eng., Vol. SE-12, No. 6 (June 1986)
9. German, R., Mitzlaff, J.: Transient Analysis of Deterministic and Stochastic Petri Nets with TimeNET. in: Beilner, H., Bause, F. (eds.), Quantitative Evaluation of Computing and Communication Systems, Lecture Notes in Computer Science 977, Springer (1995) 209–223
10. German, R., Kelling, Ch., Zimmermann, A., Hommel, G.: TimeNET - A Toolkit for Evaluating Stochastic Petri Nets with Non-Exponential Firing Times. Journal of Performance Evaluation, Elsevier, Vol. 24 (1995) 69–87
11. Heck, E., Hogrefe, D., Müller-Clostermann, B.: Hierarchical Performance Evaluation Based on Formally Specified Communication Protocols. IEEE Trans. on Comp., Special Issue on Protocol Engineering, Vol. 40 (1991) 500–513
12. Hermann, Ch.: How Good is Stationary Analysis for the Transient Phenomena of Connection Admission in ATM? in: Beilner, H., Bause, F. (eds.), Quantitative Evaluation of Computing and Communication Systems, Lecture Notes in Computer Science 977, Springer (1995)
13. Hemmer, H., Huth, T.P.: Evaluation of policing functions in ATM networks. [4]
14. Holtsinger, D.S., Perros, H.G.: Performance of the buffered leaky bucket policing mechanism. High Speed Communication Networks, Plenum Press (1992)
15. Kant, K.: Introduction to Computer System Performance Evaluation. McGrawHill (1992), Chap. 12 Petri Net-Based Performance Evaluation
16. Kelling, C., German, R., Zimmermann, A., Hommel, G.: TimeNET - Ein Werkzeug zur Modellierung mit zeiterweiterten Petri-Netzen (in German). it+ti, Vol. 3, Oldenbourg-Verlag (1995)

17. Kelling, Ch.: TimeNET-SIM - a Parallel Simulator for Stochastic Petri Nets. Proc. 28th Annual Simulation Symposium, Phoenix, AZ, USA (1995) 250–258

18. Kelling, Ch.: Rare Event Simulation with RESTART in a Petri Nets Simulation Environment. Proc. of the European Simulation Symposium 1995, Erlangen, Germany, 370–374

19. Kröner, H., Theimer, H., Briem, U.: Queueing Models for ATM Systems - A Comparison. ITC Specialist Seminar, Morristown (October 1990)

20. Kouvatsos, D.D.: Performance Modelling and Evaluation of ATM Networks, Chapman and Hall (1995)

21. Kurose, J.: Open Issues and Challenges in Providing Quality of Service Guarantees in High-Speed Networks. ACM SIGCOMM, Computer Communication Review, Vol. 23, No. 1 (January 1993) 6–15

22. Lemercier, M.: Peformance simulation of leaky buckets for ATM network access control. [31]

23. Lindemann, C.: DSPNexpress: a software package for the efficient solution of deterministic and stochastic Petri nets. Performance Evaluation 22 (1995) 3–21

24. Lucantoni, D.M., Neuts, M.F., Reibman, A.R.: Methods for the Performance Evaluation of VBR Video Traffic Models. IEEE/ACM Transactions on Networking, Vol. 2 (April 1994) 176–180

25. Krieger, U., Müller-Clostermann, B., Sczittnick, M.: Modelling and Analysis of Communication Systems Based on Computational Methods for Markov Chains. IEEE-Journal on Selected Areas in Communication - Issue on Computer-Aided Modeling, Analysis, and Design of Communication Networks, Vol. 8, Issue 9 (December 1990)

26. Marsan, M.A., Balbo, G., Conte, G.: A class of generalized stochastic Petri nets for the performance evaluation of multiprocessor systems. ACM Trans. Comp. Systems, Vol. 2, No. 2 (May 1984) 93–122

27. Marsan, M.A., Balbo, G., Conte, G.: Performance Models of Multiprocessor Systems. MIT Press (1986)

28. Marsan, M.A., Chiola, G.: On Petri nets with Deterministic and Exponentially distributed Firing Times. In: Rozenberg, G. (ed.), Advances in Petri Nets 1986, Lecture Notes in Computer Science 266, Springer (1987) 132–145

29. Monteiro, J.A.S., Gerla, M., Fratta, L.: Statistical Multiplexing in ATM Networks. Performance Evaluation 12, North Holland (1991) 157–167

30. Nikolaides, I., Onvural, R.O.: A Bibliography on Performance Issues in ATM Networks. ACM SIGCOMM, Computer Communication Review, Vol. 22, No. 5 (October 1992) 8–23

31. Perros, H., Pujolle, G., Takahashi, Y. (eds.): Modelling and Performance of ATM Technology. North Holland (1993)

32. Onvural, R.O.: Asynchronous Transfer Mode Networks: Performance Issues. Artech House (1994)

33. Perros, H.: High-Speed Communication Networks. Plenum Press (1992)

34. Rathgeb, E.P.: Modelling and Performance Comparison of Policing Mechanisms for ATM Networks. IEEE Journal on Selected Areas in Communications, Vol. 9, No. 3 (April 1991) 1440–1451

35. Saito, H.: Teletraffic Technologies in ATM Networks. Artech House Books (1994)

36. Villén-Altamarino, M., Villén-Altamarino, J.: RESTART: A method for accelerating rare event simulations. In: Cohen, J.W., Pack, C.D. (eds.), Queueing, Performance and Control in ATM (Proc. of ITC-13), Elsevier Science Publishers, North Holland (1991)

37. Tobagi, F.A., Dalgic, I.: Performance Evaluation of 10Base-T and 100Base-T Ethernets Carrying Multimedia Traffic, IEEE Journal on Selected Areas of Communication, Vol. 14, No. 7, September 1996, 1436–1454
38. Wang, Q., Frost, S.: Efficient Estimation of Cell Blocking Probability for ATM Systems. IEEE/ACM Trans. on Netw., Vol. 1, No. 2 (April 1993) 230–235

Parallel Algorithms for Computing Rank-Revealing QR Factorizations*

Gregorio Quintana-Ortí[1] and Enrique S. Quintana-Ortí[2]

[1] Departamento de Informática, Universidad Jaime I, Campus Penyeta Roja, 12071 Castellón, Spain, gquintan@inf.uji.es.
[2] Same address as first author; quintana@inf.uji.es.

Summary. The solution to many scientific and engineering problems requires the determination of the numerical rank of matrices. We present new parallel algorithms for computing rank-revealing QR (RRQR) factorizations of dense matrices on multicomputers, based on a serial approach developed by C. H. Bischof and G. Quintana-Ortí. The parallel implementations include the usual QR factorization with column pivoting, and a new faster approach that consists of two stages: a QR factorization with local column pivoting and a reliable rank-revealing algorithm appropriate for triangular matrices. Our parallel implementations include the BLAS-2 and BLAS-3 QR factorizations without pivoting since they are a good reference point, though they are not appropriate for rank-revealing purposes.

Experimental comparison shows considerable performance improvements of our new approach over classical rank-revealing algorithms on the platforms we used: an IBM SP2 platform and a cluster of SGI workstations.

We study the effect of the computer communication network and the processor computational power on the performance of the algorithms. In this case, as well as in many other parallel and distributed applications, the latency and bandwidth of the network are much more important than the processor computational power and, thus, these are the key factors impacting performance.

1. Introduction

The problem of the determination of the numerical rank of a matrix arises in many areas from science and engineering. In addition, this problem is very interesting because it is representative of a large set of applications from numerical linear algebra. We start by defining the problem to study. Let A be an $m \times n$ matrix (w.l.o.g. $m \geq n$) with singular values

$$\sigma_1 \geq \sigma_2 \geq \ldots \geq \sigma_n \geq 0, \tag{1.1}$$

and define the numerical rank r of A with respect to a threshold τ as follows:

$$\frac{\sigma_1}{\sigma_r} \leq \tau < \frac{\sigma_1}{\sigma_{r+1}}.$$

Also, let A have a QR factorization of the form

---

* All authors were partially supported by the Spanish CICYT Project Grant TIC96-1062-C03-03.

$$AP = QR = Q \begin{pmatrix} R_{11} & R_{12} \\ 0 & R_{22} \end{pmatrix}, \qquad (1.2)$$

where P is a permutation matrix, Q has orthonormal columns, R is upper triangular, and R_{11} is of order r. Further, let $\kappa(A)$ denote the two-norm condition number of a matrix A. We then say that (1.2) is a rank-revealing QR (RRQR) factorization of A if the following properties are satisfied:

$$\kappa(R_{11}) \approx \sigma_1/\sigma_r \text{ and } \|R_{22}\|_2 = \sigma_{max}(R_{22}) \approx \sigma_{r+1} . \qquad (1.3)$$

Whenever there is a well-determined gap in the singular-value spectrum between σ_r and σ_{r+1}, and hence the numerical rank r is well defined, the RRQR factorization (1.2) reveals the numerical rank of A by having a well-conditioned leading submatrix R_{11} and a trailing submatrix R_{22} of small norm.

The RRQR factorization is a valuable tool in numerical linear algebra because it provides accurate information about rank and numerical nullspace. Its main use arises in the solution of rank-deficient least-squares problems; for example, in geodesy [22], computer-aided design [24], nonlinear least-squares problems [32], the solution of integral equations [18], and the calculation of splines [23]. Other applications arise in beamforming [5], spectral estimation [10], regularization [27, 28, 42], and subset selection.

Serial algorithms for the reliable computation of rank-revealing factorizations have recently received considerable attention (see, for example [4, 6, 11, 12, 26, 34, 36]). However, the most common approach to computing such an RRQR factorization is the column pivoting procedure suggested by Golub [21]. This QR factorization with column pivoting (QRP) may fail to reveal the numerical rank correctly, but it is widely used due to its simplicity and practical reliability. A BLAS-1 version of the QRP was included in LINPACK [14]. A BLAS-2 version of the QRP is currently included in LAPACK [1]. A more efficient BLAS-3 version of the QRP has been recently developed by G. Quintana-Ortí, X. Sun and C. H. Bischof [37].

A different approach for computing RRQR factorizations has been developed by C. H. Bischof and G. Quintana-Ortí [8, 9]. It consists of two stages: a QR factorization with local column pivoting [2] plus an iterative postprocessing algorithm appropriate for triangular matrices. In serial platforms, this algorithm is faster due to the higher locality of local pivoting, the use of aggregated transformations, and the small overhead introduced by the iterative postprocessing algorithm.

So far, parallel implementations of orthogonal factorizations have basically focused on the usual BLAS-2 and BLAS-3 QR factorizations without pivoting and the BLAS-2 QRP [13, 15, 19]. C. H. Bischof [3] developed a parallel implementation of the QR factorization with local pivoting on 1-D topologies. This factorization is a good approximation to the RRQR but it does not guarantee to reveal the rank.

In this paper we present a new parallel algorithm for computing rank-revealing QR factorizations on multicomputers, based on the above mentioned serial two-stage approach, developed by C. H.Bischof and G. Quintana-Ortí [8, 9]. We compare the performances of our new two-stage algorithm with the most usual QR factorizations. The parallel algorithms were implemented on 2-D topologies, thus obtaining scalable methods. We study the effect of the communication and processor characteristics on performance.

The paper is structured as follows. In section 2 we briefly describe the serial algorithms for the basic numerical tools: QRP, QR with local pivoting and iterative postprocessing. In section 3 we discuss several aspects of parallel algorithms and their implementation, in particular data layout and message-passing issues. Section 4 presents the experimental results on an IBM SP2 platform and a cluster of SGI workstations. Finally, in section 5 we summarize our results.

2. Basic serial numerical tools

2.1 QR factorization with column pivoting

The basic scheme for the QRP, as proposed by Golub [21], can be described as shown in Figure 2.1, using the MATLAB notation. e_j is used to denote the j-th canonical unit vector $(0, \ldots, 0, 1, 0, \ldots, 0)^T$ of appropriate length.

Setup:
    Permutation vector: $\text{perm}(j) = j, \quad j = 1 : n$
    Column norm vector: $\text{colnorms}(j) = \|Ae_j\|_2^2, \quad j = 1 : n$
Reduction Steps:
For $j = 1 : n$
    1. **Pivoting:** Choose p such that $\text{colnorms}(p) = \max(\text{colnorms}(j : n))$
        If $(\text{colnorms}(p) == 0)$ STOP
        If $(j \neq p)$ then   % interchange
            $\text{perm}([j, p]) = \text{perm}([p, j]), A(:, [j, p]) = A(:, [p, j])$
            $\text{colnorms}([j, p]) = \text{colnorms}([p, j])$
        Endif
    2. **Reduction:** Determine a Householder matrix H_j such that
        $H_j A(j : m, j) = \pm \|A(j : m, j)\|_2 e_1$.
    3. **Matrix Update:**
        $A(j : m, j + 1 : n) = H_j A(j : m, j + 1 : n)$
    4. **Norm Downdate:**
        $\text{colnorms}(j + 1 : n) = \text{colnorms}(j + 1 : n) - A(j, j + 1 : n).^2$
Endfor

Fig. 2.1. The Traditional Algorithm for the QR Factorization with Column Pivoting

A Householder $m \times m$ matrix H is defined as follows:

$$H = I - \tau v v^T. \tag{2.1}$$

Its application to an $m \times n$ matrix A requires the following computation:

$$HA = A - \tau v v^T A.$$

LAPACK routine xGEQPF is BLAS-2 oriented: it first computes the row vector $w^T := v^T A$ using the BLAS-2 routine xGEMV for a matrix-vector product, and then applies the rank-1 update $A - \tau v w^T$ by means of BLAS-2 routine xGER.

2.2 A Block QR Factorization with Restricted Pivoting

The bulk of the computational work in QRP is performed in the matrix update stage, which relies on matrix-vector operations and rank-1 updates. However, on current cache-based architectures and parallel and distributed computers (ranging from personal computers to workstations and supercomputers) matrix-matrix operations perform much better. Matrix-matrix operations are exploited by using the so-called block algorithms, whose top-level unit of computation is matrix blocks instead of vectors. Such algorithms play a central role, for example, in the LAPACK implementations [1]. To arrive at a block QR factorization algorithm, we would like to avoid updating part of A until several Householder transformations have been computed. LAPACK employs the so-called compact WY representation of products of Householder matrices [40], which expresses the product

$$Q = H_1 H_2 \cdots H_{nb} \tag{2.2}$$

of a series of $m \times m$ Householder matrices (2.1) as:

$$Q = I - Y T Y^T, \tag{2.3}$$

where Y is an $m \times nb$ matrix and T is an $nb \times nb$ upper triangular matrix. Then, Q can be applied to the matrix of A by means of BLAS-3 operations:

$$QA = A - Y T Y^T A. \tag{2.4}$$

This strategy is difficult to implement in QRP with traditional pivoting, since we must update the vector *colnorms* before we can choose the next pivot column. Recently, a successful implementation of the QRP by Quintana-Ortí, Sun, and Bischof shown that it is possible to perform the matrix update by means of BLAS-3 kernels [37]. The new subroutine has been named xGEQP3 and it is included in LAPACK release 3.0.

A different approach, which is considered in this paper, is to limit the scope of pivoting as suggested by Bischof [2]. The idea is graphically depicted in Figure 2.2.

At a given stage we are done with the columns to the left of the pivot window. We then try to select the next pivot column exclusively from the

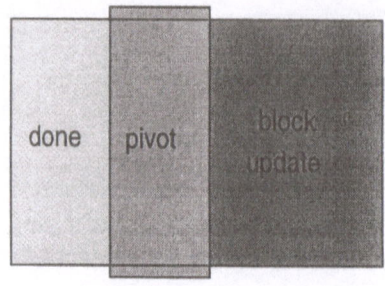

Fig. 2.2. Restricting Pivoting of the QR factorization with local pivoting. All the pivoting is usually restricted to the window labeled "pivot"

columns in the pivot window, not touching the part of the matrix to the right of the pivot window. Only when we have combined the Householder vectors defined by the next batch of pivot columns into a compact YTY^T factor do we apply this block update to the columns on the right.

The restricted block pivoting algorithm proceeds in four phases:

Phase 1: Pivoting of column with largest two-norm into first position.

The column with largest two-norm in the full matrix is pivoted to the first position. This stage is motivated by the fact that the largest two-norm column of A is usually a good estimate for $\sigma_1(A)$.

Phase 2: Block QR factorization with restricted pivoting.

Given a desired block size nb and a window size ws, $ws \geq nb$, the goal is to generate nb Householder transformations by applying the Golub pivoting strategy *only to the columns in the pivot window*, using an incremental condition estimator [7] to assess the impact of a column selection on the condition number. When the pivot column chosen from the pivot window would lead to a leading triangular factor whose condition number exceeds τ, *all* remaining columns in the pivot window are marked as "rejected" and moved to the end of the matrix. Next, a block transformation is generated, applied to the remainder of the matrix, and then the pivot window is repositioned to encompass the next ws not yet rejected columns.

In their experiments, C. H. Bischof and G. Quintana-Ortí chose the value:

$$ws = nb + \max\{10, \frac{nb}{2} + 0.05n\}. \tag{2.5}$$

Phase 3: Traditional pivoting strategy among "rejected" columns.

Since phase 2 rejects all remaining columns in the pivot window when the pivot candidate is rejected, a column may have been pivoted to the end that should not have been rejected. Hence, the traditional Golub pivoting strategy is applied on the remaining columns.

Phase 4: Block QR factorization on final columns.

The not-yet-factored columns are with great probability linearly dependent on the previous ones, since they have been rejected in both phase 2 and phase 3. Hence, it is unlikely that any kind of column exchanges among the remaining columns would change the rank estimate, and the standard BLAS-3 block QR factorization, as implemented in the LA-PACK routine xGEQRF, is the fastest way to complete the triangulariza-tion.

The factorization computed with this algorithm provides a good approx-imation to an RRQR factorization. However, this QR factorization *does not guarantee* to reveal the numerical rank correctly and a postprocessing stage is required.

2.3 Postprocessing algorithm for an approximate RRQR factorization

Chandrasekaran and Ipsen developed an algorithm, which we choose as basis of our implementation, with the following bounds in (1.2):

$$\sigma_{\min}(R_{11}) \geq \frac{1}{\sqrt{k(n-k+1)}}\sigma_k(A) \tag{2.6}$$

$$\sigma_{\max}(R_{22}) \leq \sqrt{(k+1)(n-k)}\sigma_{k+1}(A), \tag{2.7}$$

For an overview of other RRQR algorithms, see [8].

Their algorithm assumes that the initial matrix is triangular and thus it is appropriate as a postprocessing stage to the algorithm presented in the preceding section.

Bischof and Quintana-Ortí [8] presented a variant of this algorithm, whose pseudocode is shown in Figures 2.3–5.

Algorithm Hybrid-III-sf(f,k)
1. $\Pi = I$
2. **repeat**
3.   Golub-I-sf(f,k)
4.   Golub-I-sf(f,k+1)
5.   Chan-II-sf(f,k+1)
6.   Chan-II-sf(f,k)
7. **until** none of the four subalgorithms modified the column ordering

Fig. 2.3. Variant of Chandrasekaran/Ipsen Hybrid-III algorithm

---

Algorithm Golub-I-sf(f,k)
1. Find smallest index j, $k \leq j \leq n$, such that
2. $\|R(k:j,j)\|_2 = \max_{k \leq i \leq n}\|R(k:i,i)\|_2$
3. **if** $f \cdot \|R(k:j,j)\|_2 > |R(k,k)|$ **then**
4. $R := \text{triu}(R \cdot \Pi^R_{k,j}); \Pi := \Pi \cdot \Pi^R_{k,j}$
5. **end if**

Fig. 2.4. "f-factor" Variant of Golub-I Algorithm

```
Algorithm Chan-II-sf(k)
1. v := right singular vector corresponding to σ_min(R(1:k, 1:k)).
2. Find largest index j, 1 ≤ j ≤ k, such that: |v_j| = max_{1≤i≤k} |v_i|
3. if f · |v_j| > |v_k| then
4.     R := triu(R · Π^L_{j,k}); Π := Π · Π^L_{j,k}
5. end if
```

Fig. 2.5. "f-factor" Variant of Chan-II Algorithm

Compared with the original Hybrid-III algorithm, the new one incorporates the following improvements:

1. Chan-II strategy (an $O(k^2)$ algorithm) is used instead of the so-called Stewart-II strategy (an $O(k^3)$ algorithm).
2. A different loop ordering is proposed that is simpler and enhances convergence.
3. A generalization of the f-factor technique is introduced to guarantee termination in the presence of rounding errors. The analysis in [38] shows that the new variant achieves the following bounds:

$$\sigma_{\min}(R_{11}) \geq \frac{f^2}{\sqrt{k(n-k+1)}}\sigma_k(A), \text{ and} \tag{2.8}$$

$$\sigma_{\max}(R_{22}) \leq \frac{\sqrt{(k+1)(n-k)}}{f^2}\sigma_{k+1}(A), \tag{2.9}$$

where $0 < f \leq 1$. These bounds are very similar to (2.6) and (2.7), except that the factor f^2 enters the equations. We have used $f = 0.5$ in our experiments.

3. Parallelization of RRQR factorizations

3.1 Communication and parallel libraries

MPI (Message-Passing Interface) [25] is a standard high-level communication library with the following features: communication between processes through message-passing paradigm, management of message buffers, use of families of messages, declaration and management of groups of processes, use of communicators in collective communications, different communication modes, execution in heterogeneous networks, etc.

MPI reinforces portability and it is currently available on a wide range of parallel and distributed architectures like message-passing multicomputers (IBM SP2, etc.), virtual shared-memory multicomputers (Cray T3D, etc.), shared memory multiprocessors (Silicon, SUN, etc.), and clusters of workstations (SUN, HP, SGI, IBM, Linux, etc.).

We have used most of the features/capabilities of MPI to design and implement our parallel algorithms. Concretely, we employed the ANL/MSU MPI implementation (http://www.mcs.anl.gov/mpi).

Currently, there are two parallel numerical libraries under construction: ScaLAPACK [39] and PLAPACK [35]. As the former was being rebuilt and the latter was being built from scratch, we decided not to use any of them. So we developed all the parallel code we needed (parallel QR with BLAS-2, parallel QR with BLAS-3, parallel QR with pivoting, etc.) and we made extensive use of (serial) LAPACK and BLAS libraries. Thus the comparison of the new algorithms and the former ones will be fairer since all of them have been built in the same way. Finally, we decided to employ the same data distribution as ScaLAPACK since it has been successfully applied to Gaussian elimination and other similar problems.

3.2 Data layout and topology

In parallel and distributed applications, the distribution (layout) of the data among the local memories of the processors strongly affects the communication overhead and, therefore, the overall performance. In numerical problems that involve matrices, the layout of the matrices is the key factor to obtain a good performance. Thus, the data layout should achieve the following goals: balance the amount of work performed in each processor, minimize the communication overhead, and reduce the quantity of workspace.

Typically, in numerical linear algebra the assignment of the matrix elements is done at some moment in the beginning. In fact, the layout is usually "inherited" from previous computation stages and it is not modified because of the high communication cost. The pivoting is physically performed and the data is really moved between processors in order to keep the load and storage balanced.

Basic 1-D layouts such as block column/row wrap are specially appropriate for simple matrix algebra kernels which operate on small dense matrices, e.g., LU and QR factorizations, triangular linear systems, etc. [29, 31, 33]. Specialized data layouts (block Hankel wrap, etc.) have shown considerable improvements on problems that require the application of similarity transformations like the reduction to condensed forms in the eigenvalue problem [30].

On the other hand, in the last years the 2-D block wrap data layout has been applied to solve many matrix algebra problems. This data layout is generally scalable, i.e., it is possible to maintain the performance of parallel algorithms as long as the ratio *size of the problem/number of processors* holds constant [16, 17]. Hence, 2-D data layouts are usually more efficient than 1-D data layouts for large scale problems. However, programming parallel and distributed applications on 2-D layouts results much more difficult since they include column and row layouts.

Therefore, despite its more complex programming, we have used a 2-D block wrap data layout to develop our parallel algorithms. In this layout, the matrices are divided in blocks of dimension $r \times s$, and the blocks are distributed and stored locally in a $p \times q$ processor mesh of processors (see Figure 3.1). Another advantage of 2-D layout is that they include the usual 1-D layouts. For example, the 1-D column cyclic layout is a particular case of 2-D layout with $r = m$ and $s = 1$. We have performed special efforts to minimize the overhead in these particular cases.

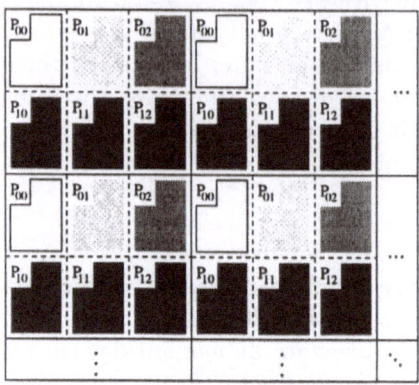

Fig. 3.1. Data layout in a mesh of 2×3 processors

This data layout defines the communication patterns and, therefore, the parallel algorithms. These algorithms are just a direct (but not easy and fast) parallelization of the serial algorithms on the 2-D layout.

This data layout has been used successfully on LU, Gaussian elimination, etc. In these cases, the processor meshes that gave the best results were $p \times q$ meshes with $q > p > 1$.

3.3 Performance parameter settings

In the second stage of the serial QR factorization with restricted pivoting presented by Bischof and Quintana-Ortí, the window size was:

$$ws = nb + \max\{10, \frac{nb}{2} + 0.05n\}. \tag{3.1}$$

On serial systems, a too large window size would make the data not fit into the cache memory. On parallel systems, it is even worse. A too large window size would make most processors be idle: while one processor is factoring the window, all the others are waiting for that one to finish. Therefore, to improve the parallelism and the performance of the application we reduced the window size:

$$ws = nb + \max\{10, \frac{nb}{2} + 0.01n\}. \tag{3.2}$$

The experimental study showed that this size achieves much better performances while not changing the numerical behavior of the algorithm.

3.4 Architecture issues

We have used and evaluated two different parallel architectures: an IBM SP2 platform and a cluster of Silicon workstations.

The IBM SP2 is a distributed-memory multiprocessor or multicomputer. Each processor is an IBM RS/6000 POWER2-370. Its main memory is 128 Mbytes, its data cache memory is 16 Kbytes, its instruction cache memory is 16 Kbytes, and its peak performance is 125 Mflops. The processors are connected through a High-Performance Switch with an Omega network. The latency of the network is about 35 microseconds and the bandwidth is about 35 Mbytes/sec.

The cluster of SGI workstations consists of 4 computers equipped with superscalar RISC processors MIPS R8000 at 75 MHz. The peak speed of each processor is 300 Mflops and it has a 4-Mbyte data cache memory.

4. Experimental results

We report in this section the experimental results comparing the following double precision parallel codes:

- DPGEQR2: BLAS-2 QR factorization.
- DPGEQRF: BLAS-3 QR factorization.
- DPGEQPF: BLAS-2 QRP factorization.
- DPGEQPB: BLAS-3 QR factorization with local pivoting.
- DPGEQPX: New two-stage RRQR factorization.

These algorithms were evaluated on an IBM SP2 platform and a cluster of SGI workstations. In both cases, we used the vendor-supplied BLAS and the ANL/MSU MPI.

We use 18 matrix types to test the algorithms, with different singular value distributions and numerical rank ranging from 4 to full rank. The matrix collection was constructed to exercise column pivoting and we expect this collection to be representative of realistic pivoting behavior. A detailed description of the matrix set can be found in [8].

The average of the results are presented in the figures.

We present results on 400 × 400 and 2048 × 2048 matrices, using block sizes (nb) 1, 8, 10, 16, 20, and 30. Topologies of 1 × 4 processors were tested on the IBM SP2 and the SGI cluster. Topologies of 1 × 16, 2 × 8, 1 × 32, and 2 × 16 were also tested on the IBM SP2.

The first experiment is designed to analyze the effect of the block size on the performance of the parallel algorithms and, also, to compare the performances of the different algorithms. Figure 4.1 shows the average performance (Mflops or millions of floating point operations per second) of the parallel algorithms on 18 different 400 × 400 matrices. An 1 × 4 topology is used both

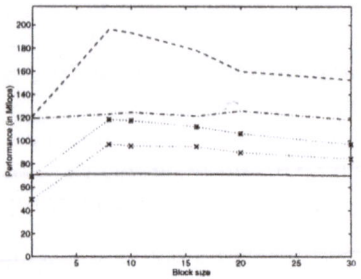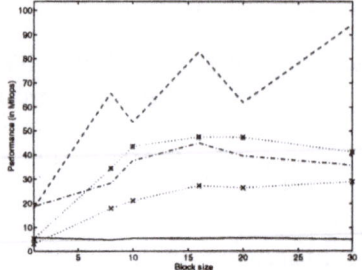

Fig. 4.1. Average Megaflops versus blocksize on the IBM SP2 (left) and on the SGI cluster (right). Solid line is employed for DPGEQPF; dotted line with symbol "*" for DPGEQPB; dotted line with symbol "x" for DPGEQPX; dashed-dotted line for DPGEQR2; and dashed line for DPGEQRF.

on the IBM SP2 and the SGI cluster. Thus, the block size is $400 \times s$, with $s = 1, 8, \ldots, 30$, i.e., a cyclic column block data layout.

The results in both platforms show that the highest performances are obtained, as we expected, by the QR factorization procedures, DPGEQRF and DPGEQR2. We are only interested in the performance of these routines because they present an optimal behavior. However, it is important to remember that these routines are not appropriate for rank-revealing purposes.

On the other hand, the lowest performance is achieved by the well-known QRP procedure DPGEQPF. This is a BLAS-2 routine and therefore is almost unaffected by the block size on both architectures. Finally, the figures show that our new two-stage parallel routine DPGEQPX performs better than the QRP procedure: 34% faster on the IBM SP2 and almost 5 times faster on the SGI cluster. The difference between DPGEQPX and the parallel QR factorization with local column pivoting DPGEQPB allows to measure the overhead of the postprocessing step.

The next experiment is designed to test the relation between the performance and the matrix type. Figure 4.2 shows the results on the SGI cluster using the same parameters (topology, block size, etc.) as in the preceding experiment. This figure compares only the performance of the parallel rank-revealing algorithms DPGEQPF and DPGEQPX. Matrix type 1 is designed to produce a failure in the preprocessing stage. Matrix types 4, 15 and 16 do not have a well-defined rank and therefore RRQR algorithms should not be applied.

Our last experiment is designed to evaluate the scalability of the parallel algorithms on the IBM SP2 platform. Figures 4.3 and 4.4 compare the execution time (in seconds) of the parallel algorithms on matrices of size 2048×2048. In the experiments, we have employed 16 processors arranged in 1×16 and 2×8 meshes, and 32 processors arranged as 1×32 and 2×16 meshes. The blocks of the distribution are square and the block sizes which

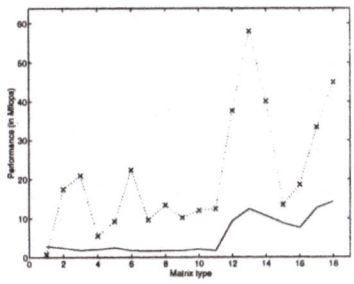

Fig. 4.2. Megaflops versus matrix type on the SGI cluster Solid line is employed for DPGEQPF; dotted line with symbol "x" for DPGEQPX.

have been tested are 1, 8, 10, 12, 16, 20, 32, and 50. Unfortunately, we have not been able to test larger matrices on these meshes. Other tested meshes such as 4 × 4 and 4 × 8 did not offer good performances. A closely similar behavior is also present in the LU factorization, but in this case it is more visible since QR factorizations require more columnwise communications.

The results in both figures show that the block size has a small effect on the performance of BLAS-2 parallel routine DPGEQR2. As was expected, Figure 4.3 shows that the behavior of DPGEQRF is completely different. The best performance of this routine is achieved with blocks sizes 8, 10, and 16. The use of 2-D topologies (2 × 8 and 2 × 16) improves the performances of the parallel algorithms almost in all cases. Larger matrices would increase the difference.

Figure 4.4 compares the execution time of the parallel rank-revealing algorithms. This figure shows that, almost for any block size, our new parallel RRQR algorithm DPGEQPX presents a lower execution time than the usual rank-revealing approach DPGEQPF. Both algorithms present better results when 2-D topologies are employed.

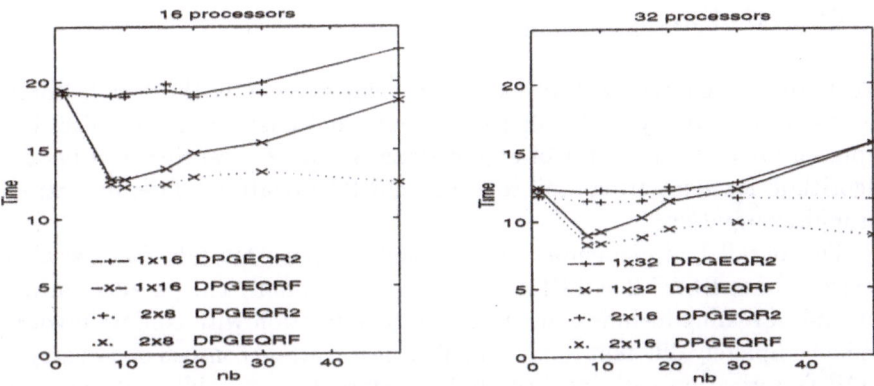

Fig. 4.3. Execution time (sec.) of the parallel QR factorizations on the IBM SP2

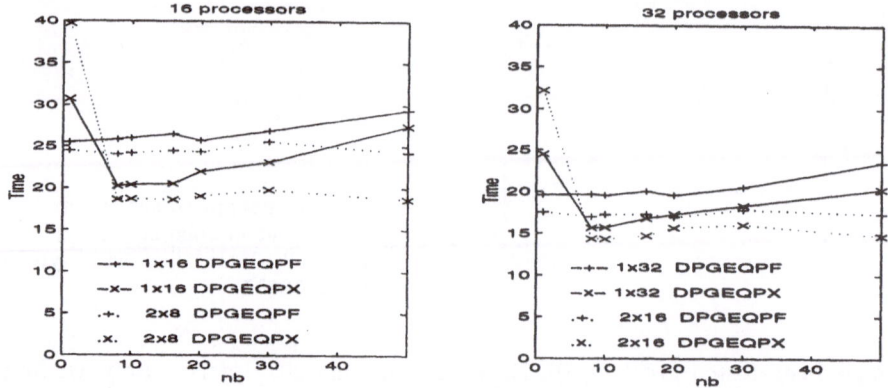

Fig. 4.4. Execution time (sec.) of the parallel RRQR factorizations on the IBM SP2

We have performed several tests on our parallel algorithms ($\|AP-QR\|_F$, $\|Q^TQ-I\|_F$, etc.). The results show that there is no significant difference between the accuracy of the serial and parallel algorithms. A thorough study of the precision of rank-revealing QR factorizations is presented in [8].

From the first experiments it is also possible to compare the performance of both computer architectures. The IBM system has slower processors (125-Mflop peak performance) than the SGI (300-Mflop peak performance); nevertheless, the former has a faster network (High-Performance Switch) than the SGI cluster (TCP/IP connections with Ethernet). Comparing the obtained experimental performance, the IBM is about 3 times faster than the SGI despite the former is about 2.4 times slower. This shows the great importance of the computer communications in parallel and distributed applications.

5. Conclusions

We have studied the rank-revealing problem since it is the key factor to the solution of many problems from science and engineering. Besides, it is representative of a large set of applications. We have presented new parallel algorithms for computing rank-revealing QR factorizations of dense matrices on multicomputers.

The parallel implementations included routines for QR factorizations based on both BLAS-2 and BLAS-3 as a reference point. The parallel routines for rank-revealing included the usual QR factorization with column pivoting and a completely different approach. This new method consists of two stages: a QR factorization with local column pivoting plus a reliable rank-revealing algorithm appropriate for triangular matrices. This approach allows the use of Householder block reflectors and therefore should perform better on parallel computers. The new parallel method is based on a serial method by Bischof and Quintana-Ortí [8].

The experimental results on the parallel platforms show that our new algorithm performs better than the usual rank-revealing approach based on the QR factorization with column pivoting. Furthermore, the use of 2-D topologies improves the experimental results for large scale problems.

We compared the performance on two different computer systems: the IBM SP2 (slower processors and a faster network), and the SGI (very fast processors and a slow network). The results show that the communication network is the key factor in this problem, as in many other parallel and distributed applications. The theoretical peak performance of the IBM is 2.4 times slower than the SGI, but our experimental results show that the IBM performs 3 times faster.

Acknowledgments

We express our gratitude to Christian H. Bischof and Xiaobai Sun for their interesting suggestions, ideas, and discussions.

Bibliography

1. E. ANDERSON, Z. BAI, C. H. BISCHOF, J. DEMMEL, J. DONGARRA, J. DUCROZ, A. GREENBAUM, S. HAMMARLING, A. MCKENNEY, S. OSTROUCHOV, AND D. SORENSEN, *LAPACK User's Guide Release 2.0*, SIAM, Philadelphia, 1994.
2. C. H. BISCHOF, *A block QR factorization algorithm using restricted pivoting*, in Proceedings SUPERCOMPUTING '89, Baltimore, Md., 1989, ACM Press, pp. 248–256.
3. ——, *A parallel QR factorization algorithm with controlled local pivoting*, SIAM J. on Scientific and Statistical Computing, 12 (1991), pp. 36–57.
4. C. H. BISCHOF AND P. C. HANSEN, *Structure-Preserving and Rank-Revealing QR Factorizations*, SIAM J. on Scientific and Statistical Computing, 12 (1989), pp. 1332–1350.
5. C. H. BISCHOF AND G. SHROFF, *On updating signal subspaces*, IEEE Trans. on Signal Processing, 40 (1992), pp. 96–105.
6. C. H. BISCHOF AND P. C. HANSEN, *A Block Algorithm for Computing Rank-Revealing QR Factorizations*, Numerical Algorithms, 2 (1992), pp. 371-392.
7. C. H. BISCHOF AND P. T. P. TANG, *A robust incremental condition scheme*, Argonne Preprint MCS-P225-0391, Mathematics and Computer Science Division, Argonne National Laboratory, 1991.
8. C. H. BISCHOF AND G. QUINTANA-ORTÍ *Computing rank-revealing QR factorizations of dense matrices*, Argonne Preprint MCS-P559-0196, Mathematics and Computer Science Division, Argonne National Laboratory, 1996.
9. ——, *Codes for rank-revealing QR factorizations of dense matrices*, Argonne Preprint MCS-P560-0196, Mathematics and Computer Science Division, Argonne National Laboratory, 1996.

10. S. F. HSIEH, J. R. LIU, AND K. YAO, *Comparisons of Truncated QR and SVD methods for AR spectral estimations*, in Proceedings SVD and Signal Processing II, 1991, Elsevier Science Publishers, pp. 403–418.

11. T. F. CHAN, *Rank-Revealing QR Factorizations*, Linear Algebra & Appl., 88/89 (1987), pp. 67–82.

12. S. CHANDRASEKARAN AND I. IPSEN, *On rank-revealing QR factorizations*, SIAM J. on Matrix Analysis and Applications, 15 (1994), pp. 592–622.

13. M. COSNARD, J. M. MULLER, AND Y. ROBERT, *Parallel QR decomposition of a rectangular matrix*, Numerische Mathematik, 48 (1986), pp. 239–250.

14. J. J. DONGARRA, J. R. BUNCH, C. B. MOLER, AND G. W. STEWART, *LIN-PACK Users' Guide*, SIAM Press, Philadelphia, 1979.

15. J. J. DONGARRA, A. SAMEH, AND D. SORENSEN, *Implementation of some concurrent algorithms for matrix factorization*, Parallel Computing, 3 (1986), pp. 25–34.

16. J. J. DONGARRA, R. VAN DE GEIJN, AND R. WHALEY, *Two dimensional basic linear algebra communication subprograms*. Computer Science Dept. Technical Report CS-91-138, University of Tennessee, 1991 (LAPACK Working Note #37).

17. J. J. DONGARRA AND D. WALKER, *The design of linear algebra libraries for high performance computers*. Computer Science Dept. Technical Report CS-93-188, University of Tennessee, 1993 (LAPACK Working Note #58).

18. L. ELDÉN AND R. SCHREIBER, *An application of systolic arrays to linear discrete ill-posed problems*, SIAM J. on Scientific and Statistical Computing, 7 (1986), pp. 892–903.

19. L. ELDÉN, *A parallel QR decomposition algorithm*, Report LiTh Mat R 1988-02, Dept. of Math., Linköping University, Sweden, 1988.

20. G. H. GOLUB AND C. F. VAN LOAN, *Matrix Computations*, The Johns Hopkins University Press, Baltimore, 2nd ed., 1989.

21. G. H. GOLUB, *Numerical methods for solving linear least squares problems*, Numerische Mathematik, 7 (1965), pp. 206–216.

22. G. H. GOLUB, P. MANNEBACK, AND P. L. TOINT, *A comparison between some direct and iterative methods for certain large scale geodetic least-squares problem*, SIAM J. on Scientific and Statistical Computing, 7 (1986), pp. 799–816.

23. T. A. GRANDINE, *An iterative method for computing multivariate C^1 piecewise polynomial interpolants*, Computer Aided Geometric Design, 4 (1987), pp. 307–319.

24. ———, *Rank deficient interpolation and optimal design: An example*, Tech. Report SCA–TR–113, Boeing Computer Services, Engineering and Scientific Services Division, February 1989.

25. W. GROPP, E. LUSK, AND A. SKJELLUM, *Using MPI: Portable parallel programming with the message-passing interface*, The MIT Press, Cambridge - Massachusetts, 1994.

26. M. GU AND S. EISENSTAT, *An efficient algorithm for computing a strong rank-revealing factorization*, Tech. Report YALEU/DCS/RR-967, Yale University, Department of Computer Science, 1994.

27. P. C. HANSEN, *Truncated SVD solutions to discrete ill-posed problems with ill-determined numerical rank*, SIAM J. on Matrix Analysis and Applications, 11 (1990), pp. 503–518.

28. P. C. HANSEN, S. TAKISHI, AND S. HIROMOTO, *The Modified Truncated SVD-Method for Regularization in General Form*, SIAM J. on Scientific and Statistical Computing, 13 (1991), pp. 1142–1150.

29. M. T. HEATH AND C. H. ROMINE. *Parallel solution of triangular systems on distributed-memory multiprocessors.* SIAM J. Scientific & Statistical Computing, 9, pp. 558-588, 1988.
30. G. HENRY AND R. VAN DE GEIJN. *Parallelizing the QR algorithm for the unsymmetric algebraic eigenvalue problem: myths and reality.* Lapack Working note #79, 1994.
31. I. C. F. IPSEN, Y. SAAD AND M. SCHULTZ. *Dense linear systems on a ring of processors.* Linear Algebra and Its Appl., 77, pp. 205-239, 1986.
32. J. MORÉ, *The Levenberg-Marquardt algorithm: Implementation and theory*, in Proceedings of the Dundee Conference on Numerical Analysis, G. A. Watson (ed.), Berlin, 1978, Springer-Verlag.
33. D. P. O'LEARY AND G. W. STEWART. *Assignment and scheduling in parallel matrix factorization.* Linear Algebra and Its Appl., 77, pp. 275-300, 1986.
34. C.-T. PAN AND P. T. P. TANG, *Bounds on singular values revealed by QR factorization*, Argonne Preprint MCS-P332-1092, Mathematics and Computer Science Division, Argonne National Laboratory, 1992.
35. A. CHTCHELKANOVA, C. EDWARDS, J. GUNNELS, G. MORROW, J. OVERFELT, R. VAN DE GEIJN *Towards Usable and Lean Parallel Linear Algebra Libraries* PLAPACK Working Note #5, TR-96-09, Department of Computer Sciences, University of Texas, May 1996.
36. G. QUINTANA-ORTÍ, *Algoritmos Secuenciales, por Bloques y Paralelos para el Cálculo del Rango Numérico Matricial*, Ph. D. Thesis, Universidad Politécnica de Valencia, 1995.
37. G. QUINTANA-ORTÍ, X. SUN, AND C. H. BISCHOF, *A BLAS-3 version of the QR factorization with column pivoting*, Argonne Preprint MCS-P551-1295, Mathematics and Computer Science Division, Argonne National Laboratory, 1995.
38. G. QUINTANA-ORTÍ AND E. S. QUINTANA-ORTÍ, *Guaranteeing termination of Chandrasekaran & Ipsen's algorithm for computing rank-revealing QR factorizations*, Argonne Preprint MCS-P564-0196, Mathematics and Computer Science Division, Argonne National Laboratory, 1996.
39. J. CHOI, J. J. DONGARRA, D. W. WALKER AND R. C. WHALEY, *ScaLAPACK Reference Manual. Parallel Factorizations Routines (LU, QR, and Cholesky) and Parallel Reduction Routines (HRD, BRD, and TRD)*, Technical Memorandum ORNL/TM-12470, Oak Ridge National Laboratory, 1994, USA.
40. R. SCHREIBER AND C. F. VAN LOAN, *A Storage Efficient WY Representation for Products of Householder Transformations*, Scientific and Statistical Computing, 10 (1989), pp. 53-57.
41. G. W. STEWART, *Introduction to Matrix Computations*, Academic Press, New York, 1973.
42. B. WALDÉN, *Using a Fast Signal Processor to Solve the Inverse Kinematic Problem with Special Emphasis on the Singularity Problem*, Ph.D. Thesis, Linköping University, Dept. of Mathematics, 1991.

A distributed algorithm for the construction of invariant subspaces

Jens Rosenboom

ABSTRACT. We describe our implementation of a distributed algorithm for the construction of subspaces of vector spaces over finite fields which are invariant under the action of one or more matrices.

1. Introduction

Numerical mathematics has developed algorithms and programs that allow for the efficient calculation with matrices which have floating point entries. These algorithms are not usable if one considers matrices over finite fields as they arise, for example, in the representation theory of finite groups. Therefore a system of Fortran programs called Meat-Axe has been developed by Richard Parker [3] to allow computations with modular representations of finite groups. A greatly enhanced version of this package has been written in C by Michael Ringe in Aachen [4] and is included as a library package in GAP [1]. By using the method of peakword condensation [2] this C-MeatAxe is able to calculate the complete submodule lattice for reasonably sized modules. However the speed and memory limitations of today's computers still make it impossible to handle representations of degree larger than some thousand in this way.

With the arrival of parallel computers the amount of memory and computing power that can be used for a single computation has increased tremendously, but of course one has to redesign the previous serial programs in order to make use of this new technology.

In this article we describe a way to implement one subtask of the Meat-Axe, the construction of invariant subspaces, on a parallel or distributed computer. The algorithm has been implemented on an IBM SP/1 using the MPC message passing library. With its help we were able to verify the results on the 2-modular character table for Conway's third sporadic simple group Co_3 which were conjectured in [7]. Because the size of the matrices was too large, Suleiman and Wilson were not able to explicitly construct a representation of degree 88000 and thereby prove the correctness of their probabilistic calculations. Our program succeeded in finding an invariant subspace of this dimension in a vector space of dimension 206184 and thus finally enabled us to verify the conjecture.

This calculation required a total amount of about 1500 cpu-hours on eight computing nodes and 2 gigabytes of main memory. A detailed description of these computations will be published separately in [5].

This work has been supported by a postdoctoral grant from the Graduiertenkolleg "Theoretische und experimentelle Methoden in der Reinen Mathematik" at the University of Essen. In particular the author would like to thank the Institute for Experimental Mathematics for providing the working environment and the computer time and resources that were needed in order to complete this project.

2. Invariant subspaces

The mathematical description of the problem that we are considering is very simple: Let $\mathbb{F} = GF(q)$ be a finite field of characteristic p and $\mathcal{A} \leq \text{Mat}(n, \mathbb{F})$ a (unitary) algebra of $n \times n$-matrices over $\mathbb{F}$ acting naturally on the vector space $V = \mathbb{F}^n$. We will assume that we are given matrices $M_1, \ldots, M_r$ generating $\mathcal{A}$. Furthermore we have a subspace $U \leq V$ given by a set of generating vectors $u^{(1)}, \ldots, u^{(k)}$ and we want to compute $U^{\mathcal{A}}$, the closure of U under the action of $\mathcal{A}$. The standard procedure for doing this is to multiply all vectors in U by all the M_i's in turn, adding newly found vectors to the original space until the dimension stabilizes.

In order to turn this procedure into an algorithm, two subtasks have to be dealt with: Multiplying vectors by a matrix and the problem of membership testing for vector spaces, i.e., given $W \subseteq V$ and $v \in V$ decide whether $v \in W$.

The first subtask can be dealt with by the matrix multiplication algorithms described in [6]. Some technical aspects arising from the special situation we have here will be discussed in section 4.

To do the membership test one uses an echelonized basis for W, i.e. a list of vectors $[w^{(1)}, \ldots, w^{(k)}]$ together with a list of *pivots* $p_1, \ldots, p_k$ such that

$$(1) \qquad w_{p_j}^{(i)} = \delta_{ij}$$

for all $1 \leq i, j \leq k$ (full echelon) resp. for all $1 \leq j \leq i \leq k$ (semi-echelon). If one has such an echelonized basis, one can test whether a vector v is contained in W by computing the sequence of vectors $v^{(0)} = v, v^{(1)}, \ldots, v^{(k)}$ defined as

$$(2) \qquad v^{(i)} = v^{(i-1)} - v_{p_i}^{(i-1)} \cdot w_i \qquad \text{for } 1 \leq i \leq k$$

Then we have $v \in W \iff v^{(k)} = 0$. Furthermore, if $v^{(k)} \neq 0$ then adding an appropriate scalar multiple of $v^{(k)}$ to $[w^{(1)}, \ldots, w^{(k)}]$ gives a semi-echelonized basis for the extended space $\langle W, v \rangle$, while to get a basis in full-echelon form one may have to change the $w^{(i)}$ by some multiples of $v^{(k)}$, too.

3. How to make use of sparsity

In this section we want to consider some performance issues involved in computing with vector spaces over small finite fields. If the size of the field is small enough one will pack several field elements into one machine word in order to allow for an efficient use of storage and arithmetic. The most extreme situation occurs for the field $\mathbb{F} = GF(2)$ where each field element can be represented by a single bit and arithmetic can be performed very efficiently by using boolean operations. The drawback of using this packing is that exploiting sparsity gets more difficult. By equation (1) an echelonized basis will have many zero entries but they have to be contigous to be visible in the packed representation. For the semi echelon form one can get the following picture by requiring that each pivot position should be the first non-zero entry of the vector and ordering the basis accordingly:

$$
\begin{matrix}
1 & * & * & \ldots & * & * & * & * & \ldots & * \\
0 & 0 & 1 & \ldots & * & * & * & * & \ldots & * \\
\vdots & \vdots & \vdots & \ddots & \vdots & \vdots & \vdots & \vdots & \vdots & \vdots \\
0 & 0 & 0 & \ldots & 0 & 1 & * & * & \ldots & * \\
0 & 0 & 0 & \ldots & 0 & 0 & 1 & * & \ldots & *
\end{matrix}
$$

So each basis vector will have zeroes up to its pivot entry and, since a pivot position cannot occur twice, we will have $p_i \geq i$, so as soon as the dimension of W exceeds our packing factor, also the packed vectors will get sparse.

To make use of the additional sparsity of a full echelon form is a bit more difficult. Using the same requirement as above, the picture will look like:

$$
\begin{matrix}
1 & * & 0 & \ldots & * & 0 & 0 & * & \ldots & * \\
0 & 0 & 1 & \ldots & * & 0 & 0 & * & \ldots & * \\
\vdots & \vdots & \vdots & \ddots & \vdots & \vdots & \vdots & \vdots & \vdots & \vdots \\
0 & 0 & 0 & \ldots & 0 & 1 & 0 & * & \ldots & * \\
0 & 0 & 0 & \ldots & 0 & 0 & 1 & * & \ldots & *
\end{matrix}
$$

The additional zeroes are interleaved with possibly non-zero entries depending on the pivot positions, so in the general case one cannot expect too much additional sparsity in the packed vectors. If we assume a random subspace, however, then

(3) $$p_{\max} \overset{\text{def}}{=} \max \left\{ i \in \{0, \ldots, k\} \,|\, \{1, \ldots, i\} \subseteq \{p_1, \ldots, p_k\} \right\}$$

will be close to k with high probability. If we then also treat the pivot entries *seperately*, *all* of the basis vectors will have zero entries up to $p_{\max}$. By our experience most real world computations behave like random data in this respect, there only few situations where $p_{\max}$ will be small. One such exception is the case when the matrices M_i have a special structure like being block matrices or a tensor product.

4. The parallel algorithm

There are different ways to convert the serial algorithm described above into a parallel program. In order to decide which way will be apropriate, one must know some data of the computer that one is going to use as well as the size of the problem that one is going to deal with. The most important thing to know is the ratio between the speed of the computation and the speed of communication. As one is usually interested in knowing the time to handle one

FIGURE 1. Example of the memory distribution

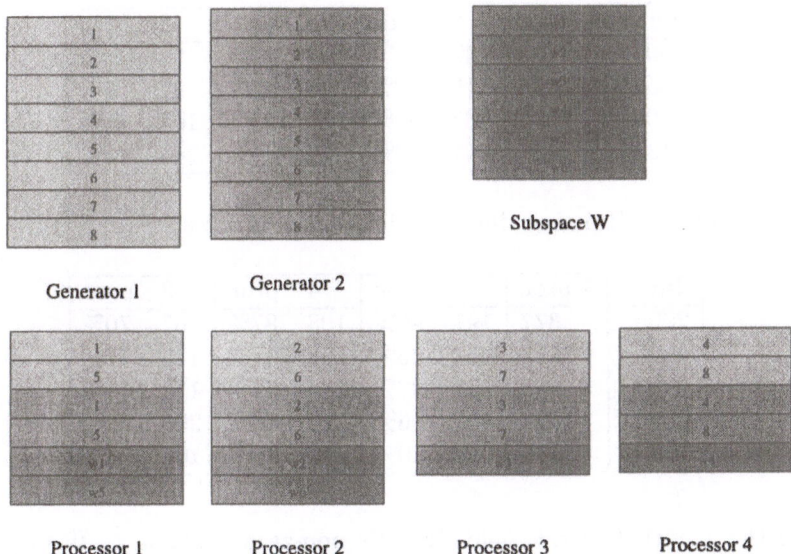

row of a matrix, this ratio also depends on the size of the problem that one wants to solve.

As our intended target machine was an IBM SP/1, which has a fast communication system, we could use the following fine granuled scheme for the computation:

The generating matrices are distributed as blocks of rows over all processors. As the subspace is constructed it will also be distributed over all processors. Figure 1 shows this type of distribution in the case of two generators and four processors.

With this configuration, the vector-matrix multiplication will be performed by broadcasting the vectors to all processors, doing all parts of the multiplication in parallel and then combining the results.

By keeping our basis B in full echelon form we have that

$$v_{p_j}^{(i)} = v_{p_j}$$

for all $i < j$, so it is possible to compute $v^{(k)}$ by having each processor compute the linear combination of his vectors which should be subtracted from v and then combining their results.

The drawback of this approach is that because of the relatively large number of broadcast and combine operations that are needed, the performance for a given problem will drop when the number of processors is increased. On the other hand, the whole running time of the algorithm is $O(n^3)$ while the amount of communication is $O(n^2)$ so that for a sufficiently large problem the communication will be neglegible.

We have implemented this algorithm in a version that was optimized for $\mathbb{F} = GF(2)$, using fast logical operations for vector additions and also making

TABLE 1. Running times for various dimensions (seconds/eff.)

dim.	1 proc	2	proc	4	proc	8	proc
4000	87	48	91%	33	66%	24	45%
5656	211	115	92%	68	78%	47	56%
8000	535	283	95%	166	81%	102	66%
11314	1486	764	97%	430	86%	244	76%

TABLE 2. Running times for different subspaces (seconds/eff.)

dim.	1 proc	2	proc	4	proc	8	proc
2000	377	197	96%	108	87%	67	70%
4000	715	371	96%	204	87%	125	71%
6000	985	508	97%	280	88%	173	71%
8000	1121	585	96%	327	86%	206	68%
10000	955	600	80%	277	86%	163	73%

use of the fact that in this field the only non-zero scalar is 1. In table 1 we give the running times for various dimensions chosen such that the size of the matrix is doubled from each row to the next. The times are given as wallclock time in seconds together with the efficiency compared to the first column.

To compare the efficiency for different sizes of subspaces we have constructed a direct sum of five irreducible modules of dimension 2000 each, and conjugated the result by a random matrix to destroy the regular structure. The times that the parallel spin algorithm needed as a function of the size of the subspace that was found are given in Table 2. Note that when our subspace gets the whole space, we know that it is invariant under any matrix, this is why the times in the last row are smaller than those above.

References

1. M. Schönert et al., *GAP — Groups, Algorithms and Programming*, Lehrstuhl D für Mathematik, Rheinisch Westfälische Technische Hochschule, Aachen, Germany, fifth edition, 1995.
2. K. Lux, J. Müller and M. Ringe, *Peakword condensation and submodule lattices: An application of the Meat-Axe*, J. Symb. Comput. **17** (1994), 529–544.
3. R. A. Parker, *The computer calculation of modular characters (The Meat-Axe)*, Computational Group Theory (Michael D. Atkinson, ed.), 1984, pp. 267–274.
4. M. Ringe, *The C-MeatAxe*, RWTH Aachen, 1994.
5. J. Rosenboom, *Verifying the 2-modular character table of Co3*, Preprint.
6. R. Staszewski, *Matrix multiplication over small finite fields on MIMD-architectures*, this issue.
7. I. A. I. Suleiman and R. A. Wilson, *The 2-modular characters of Conway's third group Co3*, J. Symb. Comput., *to appear*.

INST. FOR EXPERIMENTAL MATHEMATICS, UNIVERSITY OF ESSEN, ELLERNSTR. 29, 45326 ESSEN, GERMANY

E-mail address: jens@exp-math.uni-essen.de

ATM in Practice

Experiences in LAN and WAN Environments

Gerhard J. A. Schneider

Computing Centre, University of Karlsruhe, D-76128 Karlsruhe, Germany

1. Introduction

In recent months ATM technology has been praised as a solution to the ever rising need for bandwidth both in LAN and WAN environments, since it not only offers higher speeds at moderate cost but also the ability to assign resources according to needs. However it seems that apart from experiments in small laboratory environments little experience is available on how ATM – or rather the available ATM equipment – performs under heavy load.

The Computing Centre of the University of Karlsruhe started to analyse ATM technology with respect to data transport some time ago. Several projects finalized earlier in 1996 so that for an extended period a number of ATM equipment was available for testing, both in terms of interoperability and performance. During these tests it became clear that most ATM switches had some problems in LAN environments and were unable to handle heavy bursty traffic from many communication intensive nodes at the same time. Thus it can be questioned whether ATM is currently the right solution for high performance clusters of workstations designed to operate as a *parallel computer*.

2. Hardware environment

In 1995, in view of the pending liberalisation of the German telecommunication market, the State government of Baden-Württemberg decided to assign the contract for the high speed state university network to a consortium of two local electricity companies. ATM was chosen as the underlying technology and each of the 9 state universities will get a 155 Mbit/s link to this network, with an upgrade to higher speeds when available and necessary. The Computing Centre of the University of Karlsruhe will be responsible for the ATM network management. Since an early feasibility proof appeared necessary to all partners, a number of switches got tested at Karlsruhe in view of their suitability for a high performance network.

A number of high speed links to outside partners – three 155 Mbit/s connections and one 34 Mbit/s line – are available in Karlsruhe and therefore the tests could be carried out not only in a laboratory environment, but also over long distances. Thus problems that would have been easy to fix over small distances (and therefore underestimated) became quite apparent when the need for synchronisation of partners emerged.

At the time of the networking tests the University received its new parallel computer, an IBM RS6000/SP consisting of 100 nodes. With 82 of these nodes being 77 MHz wide nodes (28 of these nodes are owned by the Research Centre Karlsruhe) this is currently the most powerful IBM parallel computer in Europe. Since the new high performance switch for the SP was not yet available at the time of machine delivery it was decided to use ATM as an interim solution and thus to run the SP as a workstation cluster with an internal ATM network. So a large number of powerful nodes were available for testing under heavy load.

no	equipment type	#	ports	speed Mbit/s	project
2	GDC APEX	16	multi	155	CNS project
2	Newbridge Mainstreet 36150	16	multi	155	CNS project
1	FORE Switch ASX-200	4	mono	155	Campus LAN
		8	multi	155	
1	CISCO A100	4	multi	155	Campus LAN
		4	multi	100	
1	IBM 8260	24	multi	155	RS6000/SP
		4	multi	100	
2	DEC Gigaswitch				Faculty owned
1	UB Geoswitch				Faculty owned
1	HiLan ATM-Ethernet				Research Centre Karlsruhe

3. ATM — a brief overview

On an ATM network, data is sent in cells of 53 bytes, with 5 bytes being header information and a payload of 48 bytes. Such small packets should allow for low latency when passing through active networking components. In particular it is possible to assign quality of service parameters to individual connections. While many of these parameters only make sense in end to end communication, some parameter sets are useful when transporting IP packets over ATM networks. Thus ATM can be used to gradually replace the underlying networking structure of an IP network without the need to change the terminal equipment at the same time.

Currently the following service classes are used for IP networks (see [H1]):

constant bit rate	CBR
variable bit rate	VBR
unspecified bit rate	UBR
available bit rate	ABR

CBR allows to assign a constant speed to a connection - like a leased line, whereas UBR gives no guarantee of service whatsoever. In this case even loss of cells due to network congestions is accepted. ABR supplies the connected equipment with a certainty that all cells sent will actually arrive at the receiving end.

When sending IP over an ATM network, two approaches are currently used. One being *LAN emulation* (LANE), the other being *classical IP*. When classical IP is used, each host on a logical IP subnet (LIS) is directly on ATM and can connect to any other in this LIS via ATM. An *arpserver* is required to map the IP addresses onto ATM addresses (NSAP addresses).

Since ATM cells are much smaller than the typical packet size of 1500 bytes in an Ethernet, say, an IP packet has to be broken into smaller pieces, using an LLC/SNAP approach. The AAL5 mechanism used first puts an 8 byte header into the ATM payload and terminates the IP packet (which is split over many cells) with a 56 byte trailer info, containing CRC etc. This works since ATM cells always arrive in the same order as they were sent. The receiving end has to know how to read the IP packet back from the ATM cell stream. Thus traditional IP routing is not possible in this setup, nor will the application directly see any of the ATM quality of service parameters.

Connections between two end points can either be set up explicitly on a permanent basis (like operator handled phone calls) – the so-called permanent virtual circuits (PVC) – or they can be dialled on demand (switched virtual circuits or SVC). In order that SVCs can be passed through several switches, some sort of signalling between switches has to be implemented, as specified in UNI 3.0 or the newer and incompatible UNI 3.1. Unfortunately at the time of testing the various implementations were not stable enough to run in a heterogenous environment. In the meantime the situation has generally improved.

In the German Telekom ATM pilot network connections have to be ordered in advance via fax. While an early booking may be helpful, we found no problem with getting a connection within 20 minutes of calling.

The following scenario, which is currently being implemented, highlights the possibilities of sharing several fixed bandwidth connections over one high speed line and thus to make use of the degression in price for higher speeds.

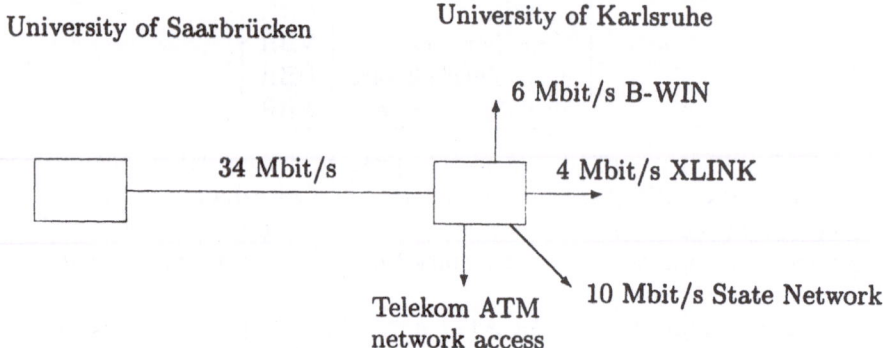

University of Saarbrücken   University of Karlsruhe

6 Mbit/s B-WIN

34 Mbit/s   4 Mbit/s XLINK

Telekom ATM   10 Mbit/s State Network
network access

4. ATM performance issues

On UNIX workstations as well as PCs the communication is handled by the CPU. Shipping large volume data at sustained high speed rates is a major challenge for system designers. In particular it remains to be seen whether modern machines can make use of the bandwidth offered by ATM.

As most applications that are currently available are based on IP communication, a number of tests were carried out measuring the throughput on standard ftp. The first round was done at Regionales Rechenzentrum der TH Hannover ([H3]).

sender		receiver	throughput
SGI challenge 2	⟶	Sun SPARC 10/41	74 Mbit/s
	⟶	SGI Onyx	87 Mbit/s
	⟶	Sun SPARC 20/51	83 Mbit/s
Sun SPARC 20/51	⟶	Sun SPARC 10/41	57 Mbit/s
	⟶	SGI Onyx	56 Mbit/s
	⟶	SGI Challenge 2	62 Mbit/s

At Karlsruhe the new 77 MHz IBM POWER2 wide node in the RS6000/SP played a major role in the tests:

sender		receiver	throughput
77 MHz POWER2	⟶	77 MHz POWER2	15.4 MByte/s
	⟶	Sun SPARC 10	3.4 MByte/s
Sun SPARC 10	⟶	Sun SPARC 10	3.0 MByte/s

In these cases the rates from disk in the sending machine to /dev/null on the receiving machine. During communication between the wide nodes, the CPU of the sending node was less than 40 % busy. This indicates that up to three adapters can be fed with data at full speed by one node. This was verified at a later stage.

Given an mtu size of 9180, 192 ATM cells are needed to transport a payload of 9132 bytes. Therefore 15.4 MByte/s require about 140 Mbit/s of available bandwidth. Since the underlying networking technology also has some management overhead, this shows that during the tests between the wide nodes traffic on the communication link reached saturation.

The tests also gave interesting insight into the architecture of the IBM ATM switch 8260. The model available in March 1996 had two 155 Mbit/s Turboways slots on one board, connected with 266 Mbit/s onto the backplane. Each of the 14 slots was used and the switch connected 28 wide nodes. There was no possibility to get port statistics from the local management, rather an installation of NetView/6000 was required.

At first SVCs were used for communication. However it turned out that the parallel IO filesystem servers on the SP were requesting and terminating SVCs at such a fast rate that the switch management could not handle them reliably, making the filesystem effectively unusable. Therefore PVCs were required, a total of 378 connections had to be entered. Unfortunately the NVRAM of the switch could only hold 110 such connections and thus after a power failure or reboot the configuration had to be reentered manually.

For flow control, IBM used the GFC bit in the ATM header. This proprietary method allows only a generic flow control at media level, but not the assignment of quality of service parameters to individual applications, such as giving preference to `telnet` over `ftp`, say. On the other hand it provided a reliable communication between the nodes.

No combination of ATM parameters however proved statisfactory to solve the problem of the typical many-to-few communication patterns (in our case, few was equal to 4) produced by the parallel fileservers. In particular CBR will not give enough bandwidth, with only $\frac{155}{24} = 6.5$ Mbit/s per connection from any node to a server, which is well below the speed of local disks. UBR will result in loss of cells under heavy load, rendering IP virtually useless. ABR is a working choice, if there is a flow control – as was available – for the combination of switch, adapters and host software.

In good cases, the four fileservers jointly managed a throughput of 28 MByte/s to disk for one parallel application.

What is really required for using ATM successfully for a high speed workstation cluster in order to turn it into a parallel computer is the ability to reduce other nodes' communication speed when necessary, so that access to critical resources can be given at full speed.

During the preparation of the transmission of a teleseminar over the ATM pilot network of the German Telekom, another ATM-related problem became apparent, when a 2 Mbit/s PVC was used. It turned out that a ping command with 56 bytes went through without problems whereas a ping with 1500 bytes resulted in a 100 % packet loss. After a rather time consuming problem hunt it became clear that the specification of 2 Mbit/s or 3622 cells per second really meant that one cell was to arrive every $\frac{1}{3622}$ second. However our equipment, which had a 155 Mbit/s adapter, was sending cells every $\frac{1}{282000}$ seconds, with an *average* throughput of 2 Mbit/s, as specified. Thus cells from a LAN switch like the FORE switch (which is able to handle LAN bursts) arrived too fast for the Telekom switch, a typical WAN switch expecting a continuous data stream. Thus it had only a four cell buffer for each port, which is sufficient for voice-type traffic. The problem could be solved by using (at that stage propietary) FORE software for IP traffic shaping.

5. Switch performance tests

Since the state network will interconnect over one hundred thousand computers, the switches for the backbone network must perform well under heavy load. In order to have reproducible performance results for individual switches the following setup was used. Two rounds of testing were carried out. In both cases the sending node was the 77 MHz IBM POWER2 wide nodes as we had proved that it was capable of operating at a maximum sustained data rate. For the receiving node we used either a SUN SPARC 10 or another wide node.

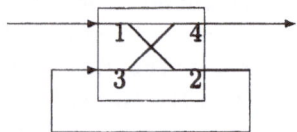

Now the following PVCs were used to have the data loop inside the switch.

$1 \longrightarrow 4$
$1 \longrightarrow 2 \longrightarrow 3 \longrightarrow 4$
$1 \longrightarrow 2 \longrightarrow 3 \longrightarrow 2 \longrightarrow 3 \longrightarrow 4$
$1 \longrightarrow 2 \longrightarrow 3 \longrightarrow 2 \longrightarrow 3 \longrightarrow 2 \longrightarrow 3 \longrightarrow 4$
$1 \longrightarrow 2 \longrightarrow 3 \longrightarrow 2 \longrightarrow 3 \longrightarrow 2 \longrightarrow 3 \longrightarrow 2 \longrightarrow 3 \longrightarrow 4$
$1 \longrightarrow 2 \longrightarrow 3 \longrightarrow 2 \longrightarrow 3 \longrightarrow 2 \longrightarrow 3 \longrightarrow 2 \longrightarrow 3 \longrightarrow 2 \longrightarrow 3 \longrightarrow 4$
... etc.

While the overall throughput can not rise above 155 Mbit/s, obviously, the loop created a rather high load on the internal switch fabric and CPU as well as on its buffer space and the buffer architecture.

The Newbridge performed rather badly. If there was more than one loop, the IP traffic collapsed completely. The switch is not able to compensate overloading via internal buffering or via other means.

With a SUN SPARC 10 as a receiving node the GDC switch could cope rather well. While the throughput to the SUN on a direct connection was only 3.4 MByte/s, 16 loops generated a throughput of 0.8 MByte/s with a load of 12.81 MByte/s on the internal switch network. The detailed results can be seen from the following plot.

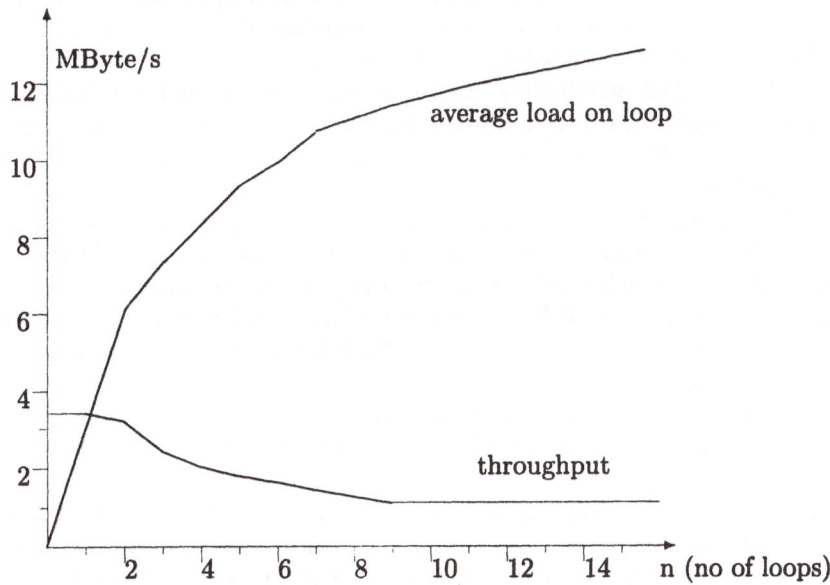

When the receiving node was replaced by another wide node, however, throughput dropped to 0.011 MByte/s with two loops only. Thus the switch started to loose cells, resulting in the usual packet loss problem on the IP level.

In contrast to these results a LAN switch like the FORE switch could cope with 15 loops or more without any negative effects on the IP communication, one reason being that it has a buffer of 32 KBytes per port.

Another interesting aspect became clear during the tests. While the GDC supports upto 5000 connections per port, the Newbridge can handle approximately 16000 connections per system.

In comparison the central University IP router currently has more than 10000 simultaneous connections at peak time. This shows that current switch technology could not support a hypothetical immediate transition of the whole Internet onto an ATM infrastructure. (Such an immediate transition is not possible for technical reasons, however.)

It is known that an ATM network requires one central clock rate. Now, what happens if one is connected to several carriers with different clock rates? Although it is possible to trigger each port of a switch externally, this may generate serious problems when one tries to do cross carrier SVCs. Fortunately in our environment the two carriers were using the same clock.

6. Management issues

When choosing a switch for a backbone network, apart from performance, management issues have to be considered as a lack of adequate management facilities may result in increased running costs.

While the GDC switch allowed for one IP address per slot, the Newbridge had the possibility of two IP addresses per switch, with only one management connection per IP address at any given time. The GDC allows one telnet session per slot.

SNMP management, a necessary condition for integrating any equipment into a professional network management system, was only possible with the GDC switch. Although PVC management appears to be rather awkward, the PVCs can be read via SNMP. On the other hand the Newbridge allows for a rather simple graphical PVC management (although the GUI does not detect preconfigured PVCs).

Available buffer space is, as seen above, essential when running IP over ATM. With the GDC, VBR is available, whereas the Newbridge has no buffers for LAN use.

During the tests we also tried to run IP with SVC connections. This was not possible with the Newbridge switch, and - although theoretically possible - did not work with our GDC. On the IBM 8230 switch, it was not only possible but it was also working. Unfortunately the switch crashed under the heavy load generated by 28 wide nodes.

Although backbone switches, especially those in a WAN environment should never be rebooted, booting time is worth noting. The Newbridge switch required 90 seconds for a reboot and the FORE switch 120 seconds. The GDC switch however required 45 seconds for a warm boot and 900 seconds for a cold boot after a power failure, for example.

Some redundancy testing was done with the Newbridge switch too. It was possible to unplug a card while the switch was in operation. Such a feature is widely available (and necessary) with WAN switches but can rarely be found in LAN switches. In fact the standby card on the Newbridge took over within 3 to 5 seconds, but failed to return control to the primary card when this card was replaced.

7. Remarks

It should be pointed out once again that these results were obtained during tests in the first quarter of 1996. Subsequent software and hardware changes may have overcome the difficulties we experienced. Yet the discrepancy between the theory of ATM and its real implementation at the time of our testing may give some interesting insight into potential problems with other manufacturers too.

These notes should not be seen as a plea against ATM. Rather, like any new technology, ATM currently still has a few pitfalls. Nevertheless, both the University LAN and the state network now are introducing ATM into its backbone infrastructure, as ATM's potential by far outweighs its deficiencies.

8. Acknowledgements

Many of the time intensive measurements were carried out by the staff of the Computer Centre. Special thanks are to W. Fries, B. Lortz, R. Steinmüller and R. Strebler.

9. References

[H1] Händel, Reiner, et al., ATM networks - concepts, protocols, applications, Addison Wesley, Wokingham, 1994
[H2] Hassenmüller, Harald, Unerwartete Stockungen, Gateway Magazin, 2/96, pp 32-37
[H3] Heerhorst, W, private communication (email, 29.6.1995)

Measurements, Proper Scaling and Application of Communication Parameters

Willi Schönauer and Hartmut Häfner

Rechenzentrum der Universität Karlsruhe
Postfach 6980, D-76128 Karlsruhe, Germany
e-mail: schonauer@rz.uni-karlsruhe.de

Abstract. The measurements of the communication parameters for different parallel computers are presented. These parameters are scaled to equivalent lost operations for startup and transfer of one operand. Then latency hiding is discussed and measured. For the example of the parallel matrix multiplication on p processors we present the theory for three different types of scaling: same problem, same storage per processor and same computation per processor. The theory is compared to measurements for the CRAY T3D, Intel Paragon, IBM RS/6000 SP with ATM-Switch and with High Performance Switch 3. Additional system overhead, compared to basic measurements, scales by a negative overlap factor.

Keywords. Parallel computing, communication, scaling, matrix multiplication

1 The Central Role of Communication

For a parallel computer communication plays a central role. Fig. 1 shows the proto-

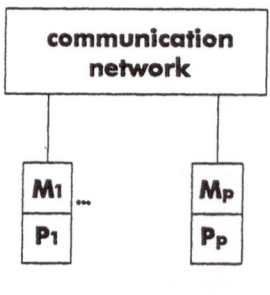

p processors

Fig. 1. Prototype distributed memory parallel computer

type distributed memory parallel computer with p processors Each of the p processors can operate only with data in its own memory. If remote data must be accessed, communication must take place by the communication network.

Some people pretend that such a computer is "scalable" to arbitrary large number p of processors. In reality scalability is limited for a given problem by the properties of the communication network, we will see this later for the example of the matrix multiplication.

Concerning communication we can distinguish three classes of "parallel" computers: workstation cluster coupled by ETHERNET, with 0.5 - 1 MB/sec transfer rate and extremely large startup time caused by the software overhead; workstation cluster coupled by special fast coupling boards like the ParaStation [1] with transfer rate in the range of 10 MB/sec and medium startup time; and finally true parallel computers with their special high speed communication network with 20 - 600 MB/sec and relatively low startup time. Between 0.5 and 600 MB/sec we have a factor of 1200 which has certainly a severe effect on the performance of the "parallel" computer.

2 Measurements and Proper Scaling of Basic Communication Parameters

The basic communication parameters are the startup time and the (asymptotic) transfer rate. We measure these parameters by the simple and double ping-pong test, see Fig. 2. We send a message of m bytes from processor A to processor B and back

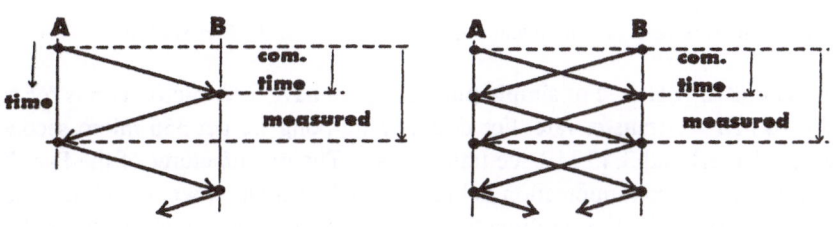

Fig. 2   Simple (left) and double (right) ping-pong test

to A, measure the time and take half of it for communication from A to B. For $m \to 0$ we get the startup time, for $m \to \infty$ we get the (asymptotic) transfer rate. If we have unidirectional communication, we get for double ping-pong half the transfer rate, if we have true bidirectional communication, we get the same transfer rate as for simple ping-pong.

As an example of such a measurement we present in Fig. 3 the printout for the ParaStation coupling of two DEC Alpha Workstations 4/275 with the Alpha 21064a

Simple ping-pong

double ping-pong

bytes	ms	MB/s	bytes	ms	MB/s
0	0.292	0.000	0	0.355	0.000
4	0.308	0.013	4	0.388	0.010
8	0.305	0.026	8	0.379	0.021
16	0.305	0.052	16	0.378	0.042
32	0.309	0.104	32	0.385	0.083
64	0.314	0.204	64	0.388	0.165
128	0.323	0.396	128	0.407	0.315
256	0.336	0.761	256	0.447	0.573
512	0.585	0.876	512	0.545	0.939
1024	0.450	2.278	1024	0.668	1.534
2048	0.561	3.649	2048	0.915	2.238
4096	0.933	4.392	4096	1.629	2.514
8192	1.546	5.298	8192	3.069	2.669
16384	2.936	5.581	16384	5.101	3.212
32768	5.693	5.756	32768	9.688	3.382
65536	10.681	6.136	65536	18.993	3.451
131072	21.270	6.162	131072	39.336	3.332
262144	43.884	5.974	262144	78.112	3.356
524288	86.294	6.076	524288	170.424	3.076

Fig. 3. Measurement of simple and double ping-pong for the ParaStation.

processor (275 MHz). For simple ping-pong we have 292 microseconds for startup and 6.16 MB/sec transfer rate. For double ping-pong we get 355 micro-seconds for startup (larger) and 3.45 MB/sec transfer rate. The manufacturer of the ParaStation pretends that the communication is bidirectional. If it were, we would get also 6.16 MB/sec transfer rate. For unidirectional communication we would get 6.16/2 = 3.08 MB/sec. So the ParaStation is close to unidirectional. The smaller values of the transfer rate for large m result from operating system overhead during the transfer time, we measure the elapsed time.

The microseconds for the startup and the MB/sec for the transfer do not tell us much about their influence on the computation. Thus we transform these parameters in $\text{lop}_{startup}$ = lost operations for one startup and in $\text{lop}_{transfer}$ = lost operations for the transfer of one 64-bit (8 B) operand [3]. For this purpose we introduce two parameters. N_{op} = number of operations that can be theoretically executed in one hardware cycle. If we have e.g. for the Power2 processor of the IBM RS/6000 Workstations two floating-point units executing with superscalar speed (simultaneous execution of addition and multiplication), we have $N_{op} = 4$.

The second parameter is the architectural efficiency η_a = real performance/theoretical peak performance for a certain operation. It tells us which part of the theoretical peak performance can be obtained as real performance for this operation. Below we present for the DEC Alpha Workstation 4/275 measurements of the vector triad

$$a_i = b_i + c_i * d_i , \tag{1}$$

which is the most important operation for engineering applications. It permits superscalar speed under the most general conditions. Unfortunately it needs four memory references, three loads and one store. The theoretical peak performance of the Alpha workstation is 275 MFLOPS for 275 MHz because it has only one floating-point pipeline for multiplication **or** addition. The measured real performance for different vector length n is

n	MFLOPS
1	19.5
10	56.7
100	60.6 → $\eta_a = 60.6/275 = 0.220$ data from cache
1000	20.6
10000	17.7
100000	6.3 → $\eta_a = 6.3/275 = 0.0229$ data from memory

The maximal performance for data in the (16 KB) first level cache is 60.6 MFLOPS, then the performance drops for data from the (2 MB) second level cache (we do not use these data), but drops to 6.3 MFLOPS for data from memory. The corresponding values of η_a are 0.220 for data from cache and 0.0229 for data from memory.

Now we can define the lost operations for startup.

$$lop_{startup} = \frac{startup\,[nsec]}{\tau[nsec]} * N_{op} * \eta_a . \tag{2}$$

Here τ [nsec] is the hardware cycle time (= 1000/f [MHz], f = frequency). The first factor are the lost hardware cycles, the first two factors are the lost arithmetic cycles (lac). The lost operations for transfer are

$$lop_{transfer} = \frac{8f \ [MB/sec]}{transfer[MB/sec]} \ * \ N_{op} \ * \ \eta_a \ . \tag{3}$$

Here transfer [MB/sec] is the measured transfer rate. 8 f is the needed transfer rate to transfer one 64 bit (8 B) operand per cycle: assume 100 MHz, then we have 100 Mwords per second or 8 x 100 MB/sec. If we measure 1/10 of that value, we need 10 hardware cycles to transfer one operand. Thus the first factor are the needed hardware cycles to transfer one operand, the first two factors are the lost arithmetic cycles (lac).

In Table 1 we present the measured communication parameters for different "parallel computers". The first three are workstation clusters connected by ETHERNET, the fourth is the ParaStation [1], a workstation cluster with special coupling boards. The IBM SP2 has the HPS2 interconnect with 40 MB/sec theoretical transfer rate, the IBM SP has the new HPS3 with 150 MB/sec. For the CRAY T3D two types of communication software have been used, the standard PVM and an own communication software based on the put/get instructions, augmented by a buffer management. One can see the extremely large startup times and low transfer rates for the workstation clusters connected by ETHERNET and the improvement by the special coupling boards of the ParaStation. In the last two columns are the lost arithmetic cycles (lac) for startup and transfer of one operand. The large values for the VPP500 result from N_{op} = 16 (8 parallel pipelines with supervector speed).

In Table 2 the lost operations for startup and transfer of one operand for the vector triad are presented. These are the values that finally are decisive, they tell us the "price" we have to pay for a startup or the transfer of one operand, measured in the currency of lost operations. These values are really instructive.

3 Latency Hiding

There is a chance to hide (part of) the communication behind the computation if
- the hardware has a separate message processor with DMA (direct memory access) that executes the communication autonomously,
- the communication software offers true asynchronous communication statements,
- the problem to be solved, its data structures and algorithm permit parallel execution of arithmetic operations and data transfer, e.g. by double buffer techniques: while one buffer is processed and sent the second buffer is filled and in the next tact the second buffer is processed and sent and the first buffer is filled. However, this must be paid by a corresponding storage overhead for the buffer(s).

Table 1. Measured communication parameters for different "parallel computers"

Parallel computer	f [MHz]	N_{op} theoret. no. of oper. for triads per hw cycle	type of comm. software	startup time simple ping pong [μsec]	max. transfer rate simple ping pong [MB/sec]	lac startup lost arithm. cycles for startup	lac transfer lost arithm. cycles for one operand
"SP/0.5" WS cluster RS6000/360	50	2	PVM	3 000	1.0	300 000	800
"SP/1.5" WS cluster RS6000/390	66.7	4	PVM	2 055	0.608	548 274	3 510
WS cluster Alpha 4/275	275	1	PVM	1 692	0.814	465 300	2 702
ParaStation Alpha 4/275	275	1	PVM (special)	292	6.16	80 300	358
IBM SP2 Thin 66 nodes	66.7	4	MPL	43	34.3	11 472	62.2
IBM SP Wide 77 nodes	77	4	MPI	51	98.3	15 700	25.0
Intel Paragon April 96	50	1	NX2	48 46	45.6 86.3	2 400 2 300	8.78 4.64
CRAY T3D	150	1	PVM	33	26.3	4 950	45.6
CRAY T3D	150	1	P-MPI Häfner	13	110.7	1 950	10.8
Fujitsu VPP 500	100	16	PAR-MACS	39	348	62 400	36.8

Table 2. Lost operations for startup and transfer of one operand for the vector triad, resulting from the communication parameters of Table 1

Parallel computer	vector triad $a_i = b_i + c_i * d_i$						
	data from cache			data from memory			
	η_a arch. eff.	lop startup	lop transfer	η_a arch. eff.	lop startup	lop transfer	
"SP/0.5" WS cluster RS6000/360	0.247	74 100	198	0.100	30 000	80	
"SP/1.5" WS cluster RS6000/390	0.252	138 165	885	0.0416	22 808	146	* same processor in WS
WS cluster Alpha 4/275	0.220	102 366	594	0.0229	10 655	62	
ParaStation Alpha 4/275	0.220	17 666	78.8	0.0229	1 839	8.20	progress by special coupling boards
IBM SP2 Thin 66 nodes	0.252	2 891	15.7	0.0416	477	2.59	*same processor in parallel computer
IBM SP Wide 77 nodes	0.432	6 786	10.8	0.150	2 356	3.8	new HPS3
Intel Paragon April 96	0.178	427 409	1.56 0.824	0.0580	139 133	0.509 0.269	better SW
CRAY T3D	0.296	1 465	13.5	0.0440	218	2.00	
CRAY T3D	0.296	577	3.20	0.0440	86	0.475	own development (Häfner)
Fujitsu VPP 500	no		cache	0.311	19 406	11.4	8 vector pipe groups

Now we want to measure latency hiding. We send a package of m bytes and determine the corresponding transfer time t_{comm}. Then we measure the computation time for a certain operation of vector length n, e.g. the vector triad (1), and determine a repeat loop count around this operation so that the computation time $t_{comp} = t_{comm}$. Then we execute a code (itself in a repeat loop) like this

> send asynchronous (start sending)
> receive asynchronous (start receiving)
> computation
> wait receive (synchronization)
> wait send   (synchronization)

and we measure the time t_{total} for one cycle.

There are two limiting cases. The ideal situation is

$t_{total} = t_{comp}$, full overlap, ol = 1.

This means that the communication has been completely hidden behind the computation, we have full overlap of communication and computation. This is expressed by an overlap factor ol = 1 (100 % overlap). The bad situation is

$t_{total} = t_{comp} + t_{comm}$, no overlap, ol = 0.

This means that no latency hiding is possible, the CPU is busy and responsible for the whole communication. This means overlap factor ol = 0 (0 % overlap).

Usually we may have the situation, that the part ol of the communication is hidden. Then we have

$$t_{total} = t_{comp} + (1 - ol)\ t_{comm}.\tag{4}$$

If we have e.g. ol = 0.4 (40 % hidable), we have 1 - ol = 0.6, i.e. 60 % of the communication is done by the CPU. This explains why we did not get other values of ol if we used in the measurements $t_{comp} = k * t_{comm}$, with k = 2, 3, ...

We made measurements with the vector triad (1) as operation. We used two types of test data. Firstly, we used a data package of m = 10 KB and the vector length of the operation n = 100. This means that the data for the computation is fixed in the cache (data from cache). Secondly, we used a data package of 800 KB and n = 10 000 (or 100 000 for large caches), so that the data does not fit in the cache (data from memory). Then there may be memory and bus contention by the message processor and the CPU.

In Table 3 we present the results of the measurements of the overlap factor ol for different "parallel computers". For the workstation clusters there is no overlap, the CPU does all the communication. For the IBM SPs there is little overlap for data from cache and no overlap for data from memory. For the Paragon, T3D and

VPP 500 there is roughly 50 % overlap, i.e. the communication processor does 50 % of the communication and the CPU 50 %.

For an "ideal" parallel computer there should be a message processor and corresponding operating system software so that ol = 1 results. Then the user has a chance to hide completely the communication behind the computation by an intelligent design of his data structures and algorithms.

Table 3. Measured values of the overlap factor ol for the vector triad for different "parallel computers"

parallel computer	type of comm. softwre	ol for data from cache	ol for data from mem.
"SP/1.5" WS cluster RS6000/390	PVM	0	0
WS cluster Alpha 4/275	PVM	0	0
ParaStation Alpha 4/275	PVM (special)	0	0.1
IBM SP2 Thin66 nodes	MPL	0.2 non-blocking 0.4 blocking	0
IBM SP Wide77 nodes	MPI	0.2	0
Intel Paragon	NX2	0.5	0.5
CRAY T3D	PVM	0.4	0.5
CRAY T3D	P-MPI Häfner	0.5	0.5
Fujitsu VPP500	Parmacs	no cache	0.5

4 Performance Prediction

If we know the parameters $lop_{startup}$, $lop_{transfer}$ and ol, we can easily predict the performance reduction by the communication overhead.

If we have a problem with **full** parallelization, i.e. no influence of Amdahl's law, no "side effects", we get, if we denote by "perf" the performance in MFLOPS or GFLOPS,

$$perf_{parallel} = perf_{single} \frac{n_{arithm}}{n_{arithm} + (st * lop_{startup} + m * lop_{transfer})(1 - ol)} . \tag{5}$$

Here $perf_{parallel}$ denotes the performance of a single processor in the parallel program and $perf_{single}$ denotes its performance as single processor, n_{arithm} denotes the number of arithmetic operations, st the number of startups and m the number of operands that must be transferred. The denominator represents the total number of operations in the parallel environment including the "equivalent" operations for startup and transfer. The fraction in (5) is a reduction factor that tells us how the performance is reduced by the communication overhead. If by the algorithm and the machine parameters complete latency hiding is possible, i.e. $ol = 1$, there is no performance reduction for the processor.

The consequences of equation (5) are rather trivial:

- avoid communication whenever possible,
- try to hide communication behind computation,
- avoid startups by sending large blocks of data,
- the manufacturers should improve the hardware and software of their parallel computers so that $lop_{startup}$, $lop_{transfer}$ are small and ol is close to one.

5 Application to Matrix Multiplication

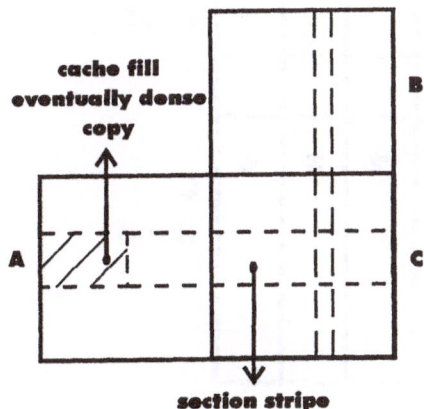

Fig. 4. Explanation for the MMUL on a single processor

The matrix multiplication (MMUL) is fully parallelizable, the only overhead is communication. An efficient parallelization starts with optimizing the algorithm on a single processor so that it is optimal for a vector processor or different types of RISC microprocessors with cache. We want to compute C = A * B, with n * n matrices A, B, C.

We give only a short description of the necessary "tricks", see Fig. 4. We use the "columnwise form" of the MMUL with "section striping" [2]: only a section of the column of C that can be kept in a vector register of a vector processor or a set of scalar registers of a RISC processor is processed, then the next section etc. For a RISC processor with cache only a cache fill of A in this section stripe is processed, i.e. the contribution of this cache fill to the whole section stripe of C. Then the next cache fill is processed etc., so that the whole section stripe of C is completed, then follows the next section stripe. If we have an IBM RS/6000 processor with a 4-way associative cache, data is dense in the cache only if it is also dense in the memory. The shaded cache fill of A in Fig. 4 is not dense in the memory. Therefore we must use a dense copy of the cache fill for such type of processor. This algorithm usually cannot be realized in Fortran because the compiler will not support the fixing of a C-section in a register (set). Therefore we use an optimized library routine. This is also used on the processors of a parallelized MMUL.

For the parallelization of the MMUL on p processors the matrices are distributed in p column blocks onto the p processors. The matrices B and C are fixed, the matrix A must be shifted in a "ring" through the p processors, this is shown in Fig. 5 for p = 4

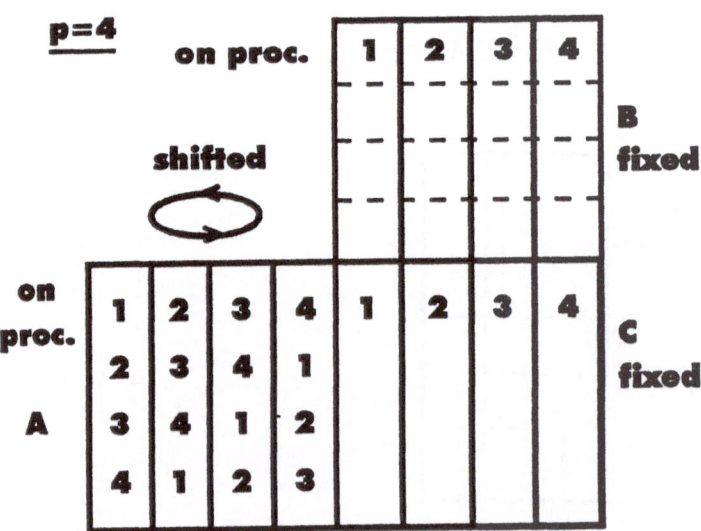

Fig. 5. Parallelization of the MMUL on p = 4 processors

processors. We use alternating buffer technique: on each processor there are two arrays a and ap for the A-block. The array a is processed and sent to the "right" processor in a ring while the array ap is received with the next block from the "left" processor. The actual column block of A is block-multiplied by the corresponding B-block and added to the C-block. This is executed by the library routine. But now there is a storage overhead of $(5 + 1/p)/3$ because instead of the 3 blocks for A, B, C we need on each processor a second buffer for A, an additional block for the intermediate block multiply by the library routine to be added to the present block of C and a dense copy of the B-block, resulting in $5 + 1/p$ blocks. Because we operate with 2 buffers and thus with double tacts, our parallel MMUL needs even p.

For the determination of the lost operations $lop_{startup}$ (2) and $lop_{transfer}$ (3) we have to use the architectural efficiency η_a for the single processor MMUL. The overlap factor ol defined in section 3 by equation (4) must be determined using double ping-pong communication and the library MMUL as arithmetic operation.

We measure the time t_1 on a single processor for matrices with the basic dimension $n_1 \times n_1$, executed by the library routine or we compute $t_1 = 2 n_1^3 / $ (MFLOPS rate, measured). For the parallelization of the MMUL we discuss 3 possiblities of "scaling":

a)   same problem solved on p processors
        (is inefficient use of parallel computer),

b)   scaling for same storage n_1^2 on each of the p processors
        (is most efficient use),

c)   scaling for same computation, $2 n_1^3$ operations, on each of the p processors
        (is medium efficient use).

Some definitons:

n_1: dimension of base problem.
t_1 : execution time for base problem.
n_p : dimension of problem for p processors, depends on type of scaling.

$$t_{p,ideal} = t_1 \; \frac{\dfrac{2n_p^3}{p}}{2n_1^3} \; , \tag{6}$$

here we have used

$$\frac{t_{p,ideal}}{t_1} = \frac{\dfrac{2n_p^3}{p}}{2n_1^3}$$

$t_{p,\,ideal}$ is the **ideal** time (no overhead) on **one** processor of the parallel computer.

$$t_p = t_1 \ \frac{\textit{"equval."oper. on one proc. of the par. comp.}}{2n_1^3} \tag{7}$$

t_p is the **real** time on one processor.

$$speedup \ sp = \frac{\textit{time to solve problem on one proc.}}{\textit{time to solve problem on p processors}} \ ,$$

$$sp = \frac{p * t_{p,ideal}}{t_p} \ . \tag{8}$$

Note that for $t_p = t_{p,\,ideal}$ we get sp = p. sp tells you, that sp of your p processors do **useful** work. p-sp processors produce overhead, thus are **lost processors!** We should rather look at p-sp than at p.

5.1 Scaling for Same Problem Solved on p Processors

The problem with fixed dimension $n_p = n_1$ is solved on p processors, see Fig. 6. The to-

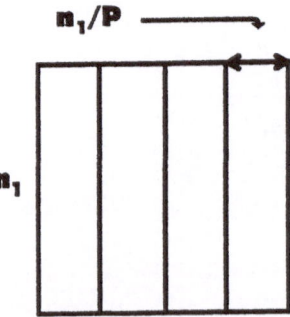

Fig. 6   Scaling for same problem solved on p processors

tal computation is $2 n_1^3 / p$, on one processor. On one processor we have p startups for the p blocks and we have to transfer n_1^2 operands. From (6) we get the ideal time on one processor of the parallel computer.

$$t_{p\,ideal} = t_1 \frac{\dfrac{2n_1^3}{p}}{2n_1^3} = \frac{t_1}{p} \quad (trivial). \tag{9}$$

The real time on one processor of the parallel computer is from (7)

$$t_p = t_1 \frac{\dfrac{2n_1^3}{p} + (p * lop_{startup} + n_1^2 * lop_{transfer})(1-ol)}{2n_1^3}. \tag{10}$$

The first term in the numerator is the computation that decreases with p , the remainder is the communication overhead that increases with p. Theoretically $t_p \to \infty$ for $p \to \infty$, but we have a limit $p \le n_1$, i.e. at least one column per processor. For $ol > 0$ we need sufficient operations to hide the communication, else $(1 - ol) \to 1$. For the speedup we get from (8)

$$sp = \frac{p * \dfrac{t_1}{p}}{t_p} = \frac{t_1}{t_p}. \tag{11}$$

Note that a timing relation like (10) is just the "inverse" of a performance relation like (5).

5.2   Scaling for Same Storage n_1^2 on Each of the p Processors

Now we have a different dimension n_p on each processor. From the condition of same storage per processor, see Fig. 7, volume of one block $n_p^2 / p = n_1^2$, we get

$$n_p = n_1 \sqrt{p}. \tag{12}$$

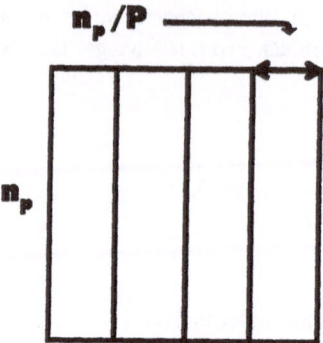

Fig. 7. Scaling for dimension n_p of the problem

The total computation is $2\,n_p^3 = 2\,n_1^3\,p\,\sqrt{p}$, thus the computation on one processor is $2\,n_1^3\,\sqrt{p}$. The ideal time is from (6)

$$t_{p,ideal} = t_1\,\frac{\dfrac{2n_p^3}{p}}{2n_1^3} = t_1 * \sqrt{p}\,.\tag{13}$$

On each processor we have p startups and $n_p^2 = p * n_1^2$ operands to be transferred. Thus from (7) we get the real time

$$t_p = t_1\,\frac{2n_1^3\sqrt{p} + (p * lop_{startup} + p * n_1^2 * lop_{transfer})\,(1-ol)}{2n_1^3}\,.\tag{14}$$

The first term in the numerator is the computation that increases now with $\sqrt{p}$, the remainder is the communication, where also the transfer term now increases with p. Theoretically $t_p \to \infty$ for $p \to \infty$, but from the condition that we have at least one column, $n_p/p = n_1/\sqrt{p} \geq 1$ we get $p \leq n_1^2$. For $ol > 0$ we need sufficient operations to hide the communication, also $(1-ol) \to 1$. From (8) we get for the speedup

$$sp = \frac{p * t_{p,ideal}}{t_p} = \frac{t_1 * p * \sqrt{p}}{t_p}\,.\tag{15}$$

5.3 Scaling for Same Computation $2\,n_1^3$ on Each of the p Processors

The dimension n_p, see Fig. 7, is now determined by the same computation per processor, $2\,n_p^3/p = 2\,n_1^3$, thus

$$n_p = n_1 p^{1/3}, \tag{16}$$

The total computation is $2 n_p^3 = 2 n_1^3 * p$, the p-fold. The total storage for a matrix is now $n_p^2 = n_1^2 p^{2/3}$, the $p^{2/3}$-fold. From (6) we get the ideal time

$$t_{p,ideal} = t_1 \frac{2 n_1^3 * p / p}{2 n_1^3} = t_1 \tag{17}$$

which is now trivially t_1. From (7) we get the real time

$$t_p = t_1 \frac{2 n_1^3 + (p * lop_{startup} + n_1^2 * p^{2/3} * lop_{transfer}) (1 - ol)}{2 n_1^3}, \tag{18}$$

which is now self-explanatory. The computation is now independent of p, this is the scaling basis. For $p \to \infty$ we have $t_p \to \infty$, but the limit of one column per processor, $n_p / p \geq 1$ restricts p to $p \leq n_1 \sqrt{n_1}$. For $ol > 0$ we need sufficient operations to hide the communication, else $(1-ol) \to 1$. From (8) we get the speedup

$$sp = p \frac{t_1}{t_p}. \tag{19}$$

5.4 Relation of Terms for Two Examples

We want to demonstrate the influence of the communication for a typical example: the same processor, an IBM RS/6000 Model 390 which is identical to the Thin 66 Node of the IBM SP or formerly called SP2. We investigate only the scaling for same problem on $p = 8$ processors. We take $n_1 = 2\,200$, then we have $t_1 = 21.3 * 10^9 / 210 * 10^6 = 101.4$ sec (we have measured 210 MFLOPS). The ideal time is $t_{8,ideal} = t_1 / 8 = 12.68$ sec.

We present below directly the terms in the t_p-formula (10). For the communication parameters we have to take the values for the MMUL.

1. Workstation cluster with ETHERNET-Switch

This is what we have named "SP 1.5" in Table 1. We get

$$t_8 = 101.4 \; \frac{\overset{t_1}{2.66} * 10^9 + \overset{2n_1^3/p}{(3.84} * 10^6 + \overset{p*lop_{st.}}{1.983} * \overset{n_1^2 lop_{tr.}}{10^{10}})(1-ol)}{\underset{2n_1^3}{21.3 * 10^9}}$$

$$= 107.1 \text{sec.}$$

So we need more time than on a single processor. The reason is the transfer term that is larger than the denominator, thus the "reduction" factor becomes an "increasing" factor: The communication overhead is larger than the mere computation.

2. SP2 with HPS2 (40 MB/sec peak)

This is the SP2 of Table 1. We get

$$t_8 = 101.4 \; \frac{\overset{t_1}{2.66} * 10^9 + \overset{2n_1^3/p}{(7.89} * 10^4 + \overset{p*lop_{st.}}{3.34} * \overset{n_1^2 lop_{tr.}}{10^8})(1-ol)}{\underset{2n_1^3}{21.3 * 10^9}}$$

$$= 14.25 \text{sec.}$$

which is not too far away from $t_{8,ideal}$. We get from (11) $sp = t_1 / t_p = 7.12$ useful processors and $p - sp = 0.88$ processors lost for communication overhead.

These two examples demonstrate drastically the difference between a workstation cluster and a "true" parallel computer.

5.5   Application to the CRAY T3D

The processor of the CRAY T3D is the Alpha 21064, 150 MHz, $\tau = 6.66$ nsec, $r_{theor} = 150$ MFLOPS, MMUL (library) 107.83 MFLOPS, $\eta_a = 107.83/150 = 0.71886$ for MMUL, number of operations per cycle $N_{op} = 1$ (no superscalar speed). The communication for double ping-pong has been measured with our own P-MPI built on put/get with startup 15 microsec, transfer 69.9 MB/sec. For the MMUL as arithmetic operation $ol = 0$. From the above values result the lost operations

$$lop_{startup} = 1\,617.45, \quad lop_{transfer} = 12.341 .$$

Scaling for same problem. We use the formulas of section 5.1. We could measure until p = 32 processors. The dimension was $n_1 = 1\,500$, the time on one processor $t_1 = 62.6$ sec. The results of the measurements and of the theory are presented in Table 4 up to p = 32, the theory is also applied to p = 64. The theory fits excellently to the measurements

Scaling for same storage per processor. We use the formulas of section 5.2. We did only two control measurements. Now we can use only $n_1 = 1250$ because of the storage overhead. The single processor time is $t_1 = 36.22$ sec. The results are presented in Table 5. Now the dimension increases with p: $n_p = n_1 \sqrt{p}$, see col. 1 in Table 5. The control measurement for p = 16 fits only moderately.

Scaling for same computation per processor
We use the formulas of section 5.3. We take $n_1 = 1\,250$, then $t_1 = 36.22$ sec. The dimension increases with p: $n_p = n_1 \times p^{1/3}$. The results are presented in Table 6.

Table 4. Results of measurements and theory for the T3D for scaling for same problem, $n_1 = 1\,500$

	timings [sec]				sp	p-sp
		measured		t_p	for	lost
				theory	④	pro-
	$t_{p,ideal}$	from	to	ol = 0		cessors
col.	①	②	③	④	⑤	⑥
p = 2	31.3	32.48	32.48	31.6	1.98	0.02
4	15.6	15.95	15.95	15.9	3.94	0.06
8	7.82	8.09	8.14	8.08	7.75	0.25
16	3.91	4.25	4.31	4.17	15.01	0.99
32	1.96	2.24	2.27	2.21	28.3	3.7
64	0.980	-	-	1.24	50.6	13.4

Discussion. If we look for p = 64 at p - sp, the lost processors for communication, we see that we lose for the three types of scaling 13.4, 2.43 and 4.68 processors. This demonstrates that scaling with same storage per processor is the best, same problem is the worst use of a parallel computer, but observe the larger dimensions n_p and correspondingly the larger timings. If you use your parallel computer that your problem is solved faster, you have to pay by lost processors for communication.

Table 5. Results of measurements and theory for the T3D for scaling for same storage per processor

		timings [sec]					
			control measurements		t_p theory ol = 0	sp for ⑤	p-sp lost pro-cessors
	n_p	$t_{p,ideal}$	from	to			
col.	①	②	③	④	⑤	⑥	⑦
p = 2	1768	51.22			51.58	1.99	0.01
4	2500	72.44	73.33	73.34	73.16	3.96	0.04
8	3536	102.45			103.88	7.89	0.11
16	5000	144.88	150.75	151.74	147.74	15.69	0.31
32	7071	204.89			210.61	31.13	0.87
64	10000	289.76			301.2	61.57	2.43

Table 6. Result of theory for the T3D for scaling for same computation per processor

		timings [sec]		sp for ③	p-sp lost pro-cessors
	n_p	$t_{p,\,ideal}$	t_p theory		
col.	①	②	③	④	⑤
p = 2	1575	36.22	36.50	1.98	0.02
4	1984	36.22	36.67	3.95	0.05
8	2 500	36.22	36.94	7.85	0.15
16	3 140	36.22	37.36	15.51	0.49
32	3 968	36.22	38.02	30.49	1.51
64	5 000	36.22	39.08	59.32	4.68

The most efficient use of a parallel computer is to fill up the memory of each processor, i.e. to use larger p only for larger problem size. In Table 5 we solve the problem for n_p = 5000 with p = 16 processors in 147.7 sec and lose 0.31 processors, whereas in Table 6 we solve the same problem for n_p = 5000 with 64 processors in 39.1 sec, but we lose 4.68 processors. This is the price for the faster solution.

5.6   Application to the Intel PARAGON

The processor of the Intel PARAGON is the i860/XP processor, 50 MHz, τ = 20 nsec, r_{theor} = 50 MFLOPS, MMUL (library) 45.85 MFLOPS, η_a = 45.85/50 = 0.917 for MMUL, N_{op} = 1 (no superscalar speed). The communication for double ping-pong with NX2 message passing yields 65 microsec for startup and 34.9 MB/sec transfer rate. The resulting lost operations are

$$lop_{startup} = 2\,980, \quad lop_{transfer} = 10.51.$$

The measurement of the overlap factor ol with the MMUL as operation, immediately showed the problemacy of ol for the PARAGON: we got values ol = -0.1 to -0.3, neither scaled with the measurements. Then we measured a ring shift (like it is used in the parallel MMUL) with p = 16 processors. We measured ol = -0.7 to - 2.6, in the mean -1.4. Negative ol means there is additional overhead in the MMUL compared to the "pure" communication measurements with "separate" MMUL in the "background". The reason is that a node of the PARAGON is in reality a SMP, because it has two i860/XP processors: one as arithmetic processor and one as message processor. On each processor are executing many UNIX threads that cause unpredictable additional overhead in the interplay of the processors in the MMUL. In section 6 we discuss further the meaning of negative ol.

Scaling for same problem. We use the Formulas of section 5.1. The dimension was n_1 = 1000, t_1 = 43.62 sec. The results of the measurements and of the theory are presented in Table 7. Because of the scattering of ol we computed t_p for different values of ol and compared the result to the measurements. It turned out that ol = -2.3 fitted well to the measurements, see Table 7.

Scaling for same storage per processor. We use the formulas of section 5.2. We did only two control measurements. Now we can use only n_1 = 750 because of storage overhead, with t_1 = 18.48 sec. The results are presented in Table 8. The dimension increases with p: $n_p = n_1 \sqrt{p}$, see col. 1 in Table 8. For this scaling now ol = -1.4 fitted the measurements.

Scaling for same computation per processor. We use the formulas of section 5.3. We take again n_1 = 750, t_1 = 18.48 sec, and we take ol = -1.4. The results are presented in Table 9. The dimension now increases with p: $n_p = n_1 * p^{1/3}$, see col 1 in Table 8.

Table 7. Results of measurements and theory for the PARAGON for scaling for same problem, $n_1 = 1000$

col.	timings [sec]				sp for ④	p-sp lost pro-cessors
	$t_{p,ideal}$	measured		t_p theory ol = - 2.3		
		from	to			
col.	①	②	③	④	⑤	⑥
p = 2	21.81	22.43	22.44	22.56	1.93	0.07
4	10.91	11.51	11.55	11.66	3.74	0.26
8	5.453	6.000	6.161	6.211	7.02	0.98
16	2.726	3.446	3.711	3.486	12.51	3.49
32	1.363	2.112	2.359	2.126	20.52	11.48
64	0.6816	1.458	1.686	1.452	30.04	33.96

Table 8. Results for measurements and theory for the Intel PARAGON for scaling for same storage per processor

col.	n_p	$t_{p,ideal}$	timings [sec]		t_p theory ol = - 1.4	sp for ⑤	p-sp lost pro-cessors
			control measurements				
			from	to			
col.	①	②	③	④	⑤	⑥	⑦
p = 2	1061	26.13			26.76	1.95	0.05
4	1500	36.96	37.71	37.89	38.20	3.87	0.13
8	2121	52.27			54.76	7.64	0.36
16	3000	73.92	77.54	78.38	78.89	14.99	1.00
32	4243	104.5			114.5	29.21	2.79
64	6000	147.8			167.7	56.41	7.59

Discussion. If we look for p = 64 at the three types of scaling, we see that we lose 33.96, 4.68, 7.59 processors for the communication overhead, but the dimensions are quite different with n = 1000, 6000, 3000, and correspondingly the timings. If we solve for n_p = 3000 with p = 16 processors, we need 78.89 sec and lose 1.01 processors, see Table 8. However, if we solve the same problem with 64 processors, we need only 23.46 sec but lose 13.55 processors, see Table 9. So 13.55 - 1.01= 12.54 lost processors are the price for 78.89 - 23.46 = 55.43 sec gain in computation time. This demonstrates drastically the chances and dangers of a parallel computer.

Table 9. Result of theory for the Intel PARAGON for same computation per processor.

	n_p	timings [sec]		sp for ③	p-sp lost pro- cessors
		$t_{p, ideal}$	t_p theory ol = - 1.4		
col.	①	②	③	④	⑤
p = 2	945	18.48	18.97	1.95	0.05
4	1190	18.48	19.26	3.84	0.16
8	1 500	18.48	19.72	7.50	0.50
16	1 890	18.48	20.46	14.45	1.55
32	2 381	18.48	21.62	27.36	4.64
64	3 000	18.48	23.46	50.41	13.59

Here we have met a computer which cannot be modelled as easily as the CRAY T3D because of its complicated node architecture and software. We shall discuss this in more detail in section 6.

5.7 Application to the IBM RS/6000 SP

The processor is the Wide 77 Node with the Power2 processor, 77 MHz, τ = 13 nsec, r_{theor} = 308 MFLOPS, N_{op} = 4 because it has double floating-point units with superscalar speed. Here we can drastically demonstrate the influence of the communication because we had at first an intermediate configuration of theSP with an ATM switch as communication network until the final High Performance Switch HPS3 could be delivered. The ATM Switch is notincluded in Table 1 and 2 because it is not standard equipment.

5.7.1 Special configuration with ATM Switch

Measurements of the MMUL with the library routine, up to n = 512, yielded 290 MFLOPS, thus η_a = 290/308 = 0.941. The ATM-Switch has theoretically 155 Mbit/sec = 19.4 MB/sec bidirectional transfer rate. We measured with MPI for double ping-pong 609 microsec for startup and 4.98 MB /sec transfer rate.Thus we get for the lost operations $lop_{startup}$ = $1.763 * 10^5$ (!) and $lop_{transfer}$ = 465. A measurement of the overlap factor ol was not possible because the processors of the intermediate configuration were used as a workstation cluster and not as a fixed partition like on a "true" parallel computer. Therefore we "assumed" ol = 0, although a negative value of ol, even on a fixed partition, would be more realistic. Thus the results of the theory are more optimistic than the reality.

Table 10. Results of the theory for the SP with ATM switch for scaling for same problem, n_1 = 3000

	timings [sec]		sp for ③	p-sp lost pro-cessors
	$t_{p, ideal}$	t_p theory ol = 0		
col.	①	②	③	④
p = 2	93.1	107.3	1.74	0.26
4	46.6	60.70	3.07	0.93
8	23.3	37.43	4.97	3.03
16	11.6	25.80	7.22	8.78
32	5.82	19.99	9.31	22.69

Scaling for same problem. We use the formulas of section 5.1. The dimension is n_1 = 3000, t_1 = 186.2 sec. The results are presented in Table 10. Because of the rather slow communication we lose for this scaling for p = 32 (no reasonable user would use more processors for such a configuration) 22.69 processors for communication overhead, only 9.31 processors do MMUL.

Scaling for same storage per processor. We use the formulas of section 5.2. We take n_1 = 3000, t_1 = 186.2 sec. The results are presented in Table 11. The dimension increases with p: $n_p = n_1 \sqrt{p}$, see col. 1 in Table 11.

Scaling for same computation per processor. We use the formuals of section 5.3. We take again $n_1 = 3000$, $t_1 = 186.2$. The results are presented in Table 12. The dimension increases with p: $n_p = n_1 * p^{1/3}$, see col. 1 in Table 12.

Discussion. At p = 32 we lose for the three types of scaling 22.69, 9.62, 13.88 processors for communication overhead, but at quite different dimensions n= 3000, 16971, 9524. Nevertheless the SP with ATM switch has still more or less the character of a workstation cluster than of a true parallel computer.

Table 11. Result of theory for the SP with ATM switch for scaling forsame storage per processor

| col. | timings [sec] | | | sp for ③ | p-sp lost pro-cessors |
	n_p	$t_{p, \text{ideal}}$	t_p theory ol = 0		
col.	①	②	③	④	⑤
p = 2	4 243	263.3	291.6	1.81	0.19
4	6 000	372.4	429.0	3.47	0.53
8	8 485	526.7	639.9	6.59	1.41
16	12 000	744.8	971.2	12.26	3.73
32	16 971	1053.3	1506.2	22.38	9.62

5.7.2 SP with HPS3

Measurements of the library routine for the MMUL at n = 3000 yielded 293 MFLOPS, thus we get $\eta_a = 293/308 = 0.9513$. The communication parameters of the HPS3 were measured with 70 microseconds for startup and 53.6 MB/sec transfer rate for double ping-pong with MPI. Thus we get for the lost operations
$$\text{lop}_{\text{startup}} = 20498, \quad \text{lop}_{\text{transfer}} = 43.73$$
which are roughly by an order of magnitude better than those for the ATM switch. We did measurements for the overlap factor with ring shift of 16 processors and MMUL as arithmetic operation and got ol = -0.2. But this value did not fit the scaling measurements.

Scaling for same problem. We use the formulas of section 5.1. The dimension was $n_1 = 3000$, $t_1 = 184.3$ sec. The results of the measurements up to p = 26 (max. possible at that time) are presented in Table 13. We present two columns (4) and (5) for the theory: col. (4) with ol = - 0.2 that we got from the ring shift measurements, but this value did not fit the measurements, and ol (5) with ol = -0.6 that fits rather well to the measurements. This means again that we have additional overhead in the true MMUL, see discussion in section 6. We have extended the theory up to p = 128.

Table 12.   Result of theory for the SP with ATM switch for scaling for same computation per processor

		timings [sec]		sp for ③	p-sp lost pro- cessors
	n_p	$t_{p, ideal}$	t_p theory ol = 0		
col.	①	②	③	④	⑤
p = 2	3 780	186.2	208.7	1.78	0.22
4	4 762	186.2	227.9	3.36	0.64
8	6 000	186.2	242.8	6.13	1.87
16	7 560	186.2	276.1	10.79	5.21
32	9 524	186.2	328.9	18.12	13.88

Scaling for same storage per processor. We use the formulas of section 5.2. We used $n_1 = 3000$, $t_p = 184.3$ sec. The results of the theory for ol = - 0.6 are presented in Table 14. Now the dimension increases with p: $n_p = n_1 \sqrt{p}$, see col. 1 in Table 14.

Scaling for same computation per processor. We use the formulas of section 5.3. We take again $n_1 = 3000$, $t_1 = 184.3$ sec. The results are presented in Table 15. The dimension now increases with p: $n_p = n_1 * p^{1/3}$, see col 1 in Table 15. We took again ol = - 0.6.

Discussion. If we look for p = 128 at the three types of scaling, we see that we lose 76.85, 14.9, 29.25 processors for the communication overhead. If we look at all the 3 tables we see how darstically the lost processors increase from 32 to 64 to 128 processors. This is not Amdahl's law but pure communication overhead.

Table 13.　　　　Results of measurements and theory for the SP with HPS3 for scaling for same problem, $n_t = 3000$

	timings [sec]					sp for ⑤	p-sp lost processors
		measured		t_p theory ol = - 0.2	t_p theory ol = - 0.6		
	$t_{p,ideal}$	from	to				
col.	①	②	③	④	⑤	⑥	⑦
p = 2	92.15	95.09	95.24	93.76	94.30	1.95	0.05
4	46.08	48.73	49.38	47.69	48.22	3.82	0.18
8	23.04	25.18	25.19	24.65	25.19	7.32	0.68
16	11.52	13.65	13.69	13.13	13.67	13.48	2.52
26	7.088	9.271	9.312	8.702	9.240	19.94	6.06
32	5.759				7.912	23.29	8.71
64	2.780				5.036	36.60	27.40
128	1.440				3.603	51.15	76.85

5.8 A Cross Comparison.

Now we have investigated 4 parallel computers: the T3D, PARAGON, SP with ATM and SP with HPS3. In Table 16 we compare for p = 32 the lost processors p-sp for the 4 different parallel computers. The last two computers clearly demonstrate the improvement by roughly a factor 10 in communication parameters from ATM to HPS3, but the SP with ATM is not standard and should no longer be discussed, it is still more or less a workstation cluster. If we then look at Table 16 for the 3 "true" parallel computers, the T3D seems to be the best design because for 32 processors the value p-sp is the smallest one. But we must also look at the dimension n of the matrices. Clearly the T3D is superior to the PARAGON because it loses less processors for a larger problem. But how to compare the T3D and the SP with HPS3? From Table 5 we can see that we can solve for n = 10 000 with p = 64 processors on the T3D and we lose 2.43 processors, wheras with the SP we can solve a problem of n = 16 791 with p = 32 processors and we lose 2.51 processors, if we fill up the memory of the processors. Similarly for scaling for same computation per processor we solve on a T3D a problem with n = 5000 on 64 processors and lose 4.68

Table 14. Result of theory for the SP with HPS3 for scaling for same storage per processor

| col. | n_p | timings [sec] | | sp for ③ | p-sp lost pro-cessors |
		$t_{p, ideal}$	t_p theory ol = - 0.6		
col.	①	②	③	④	⑤
p = 2	4 243	260.6	264.9	1.97	0.03
4	6 000	368.6	377.2	3.91	0.09
8	8 485	521.3	538.5	7.74	0.26
16	12 000	737.2	771.6	15.29	0.71
32	16 971	1 024	1 111	29.49	2.51
64	24 000	1 474	1 612	58.52	5.48
128	33 941	2 085	2 360	113.1	14.9

Table 15. Result of theory for the SP with HPS3 for same computation per processor

| col. | n_p | timings [sec] | | sp for ③ | p-sp lost pro-cessors |
		$t_{p, ideal}$	t_p theory ol = - 0.6		
col.	①	②	③	④	⑤
p = 2	3 780	184.3	187.7	1.96	0.04
4	4 762	184.3	189.7	3.89	0.11
8	6 000	184.3	192.9	7.64	0.36
16	7 560	184.3	197.9	14.90	1.10
32	9 524	184.3	206.9	28.63	3.37
64	12 000	184.3	218.7	53.93	10.07
128	15 119	184.3	238.9	90.75	29.25

Table 16. Cross comparison for p = 32 of 4 parallel computers for different scaling. Observe the different dimensions n

	p-sp, lost processors (in paranthesis dimension n)		
scaling of section	5.1	5.2	5.3
T3D	3.7 (1 500)	0.87 (7 071)	1.51 (3 968)
PARAGON	11.5 (1 000)	2.79 (4 243)	4.64 (2 381)
SP with ATM	22.7 (3 000)	9.6 (16 971)	13.9 (9 524)
SP with HPS3	8.71 (3 000)	2.51 (16 971)	3.37 (9 524)

processors whereas on the SP we solve a problem with n =9524 on 32 processors and lose 3.37 processors. So we see that a comparison is not easy, if we want to evaluate the communication. Here we meet an unexpected influence of the memory size of a node. A larger memory allows to use less processors and thus we have also less lost processors. But then one may discuss the corresponding execution time. What we have to pay for shorter execution time can be seen for the different computers by comparing for the scaling "same storage per processor" with the scaling "same computation per processor".

6 A Remark to Negative Overlap Factor ol

In section 3 we have discussed the possibility of latency hiding and we have determined an overlap factor ol by equation (4). Here we measure for the example of the MMUL the sending of a data package with double ping-pong and in parallel the execution of a MMUL library routine, but both operate on separate data. The overlap factor thus determined for the T3D was ol = 0 and this value fitted excellently to the theoretical values for t_p , see Table 4.

The problem came with the PARAGON. For the double ping-pong with MMUL as operation we got ol = - 0.1 to - 0.3 and with the double ping-pong replaced by a ring shift over 16 processors we got values ol = - 0.7 to - 2.6, in the mean - 1.4, see section 5.6. For the SP with HPS3 we got for the 16-processor ring shift with MMUL as operation ol = - 0.2. However, in both cases these values of ol did not fit the measurements of the parallel MMUL. Why?

Let us at first discuss, what means a negative value of ol. When we discussed equation (4) we had implicitly assumed that $0 \le ol \le 1$ holds. But when we did the measurements with the MMUL as "operation", we got a total time larger than the sum of the parts t_{comp} and t_{comm}. This means

$$t_{total} = t_{comp} + (1 - ol) \, t_{comm} > t_{comp} + t_{comm},$$

from which follows $ol < 0$. This means that the execution of the computation and of the communication "in parallel" (i.e. at the "same" time) causes for these computers additional overhead. This can be only overhead generated by the operating system. This is also the key explanation why the values of ol determined in this "artificial" way did not fit the measurements: in the execution of the parallel MMUL the computation and the communication operate on the same data (in the overlap measurements they did not), which causes additional overhead. This overhead comes from the extremely complicated interplay of the UNIX operating system and of the message passing software. There are many UNIX threads executing in a time slicing mode, producing overhead by context switches of the CPU. For the T3D we have only an optimized microkernel that avoids such overhead.

Is now our theory wrong? This was our first assumption. Look at Table 13 for the SP with HPS3. In col. (4) we have listed the values of t_p for the theory, using the value $ol = -0.2$ determined from the "artificial" ol-measurement. They do not fit the measurements, they are too small, in the real MMUL is more overhead. Then we "played" with the value of ol and for $ol = -0.6$ the theory fitted well the measurements.

A similar situation holds for the PARAGON. In Table 7 the theory for $ol = -2.3$ fits well the measurements. However, in Table 8, for another scaling, we had to use $ol = -1.4$ in order to fit the two control measurements. Both values are in the range $ol = -0.7$ to -2.6 that we got for repeated ol-measurements. This large scattering is a consequence of the complicated architecture of a node of the PARAGON, which has two i860/XP processors, each one running its UNIX, one for arithmetic, the other for communication.

Is now the theory true or false? If we look at equation (10), the formula for t_p for same problem, we see that it depends in a nonlinear way from the number of processors p. If we can adapt t_p for a wide range of p to the measurements by choosing the adequate value of ol, we can conclude that the theory is **true**. This means, that the additional overhead is due to the communication and is functionally dependent on the $lop_{startup}$ and $lop_{transfer}$ terms like the "directly" measured overhead. The additional operating system overhead is caused by the same effects like the primary overhead. We think that this is an important conclusion that can be drawn from these investigations.

7 Concluding Remarks

We may ask: Is there a "best" parallel computer? Usually a customer will buy a parallel computer because of its price performance relation DM/GFLOPS or $/GFLOPS and/or its price per memory relation DM/GB and/or $/GB. How to evaluate easily a parallel computer has been discussed in [3]. However, with this computer he buys also its communication network with its inherent properties as discussed in this paper.

We may also ask: What is a user-friendly parallel computer, concerning its communication properties? We consider as most important property the overlap factor ol. If we have ol close to one we can hide the communication behind the computation by a sophisticated design of the algorithm and data structures as we have seen it for the MMUL. However, for the PARAGON and the SP we met negative values of ol which means additional operating system overhead caused by the communication. This is not user-friendly. In principle ol close to one could be obtained by a careful separation of the computation and the communication. The CPU should only start the communication which is then executed completely autonomous by a message processor with DMA. Hopefully, the manufacturers will design in the future communication networks with such a property.

The other two parameters $lop_{startup}$ and $lop_{transfer}$ should be as small as possible. Their values are limited by the use hardware technology. Unfortunately, there is a large gap between the hardware and the measured properties that include the software. So the manufacturers must try to reduce the "terrible" software overhead in the communication routines.

In this paper we have discussed the application of the basic communication parameters $lop_{startup}$, $lop_{transfer}$ and ol to the parallel MMUL, where communication is the only overhead. If we have a more complicated algorithm, e.g. the Gauss algorithm or an iterative CG (conjugate gradient) type method, we can in principle proceed how we did it for the MMUL, but we have to break down the complicated algorithm into its basic parts that can be treated like the MMUL. If we have e.g, an algorithm with sequential parts, i.e. Amdahl's law plays a role, we can apply the methods of this paper to the parallel part and see how many "effective" operations result, if we include the "equivalent" operations for the communication.

References

1.   T.M. Warschko, J.M. Blum, W.S. Tichy, The ParaStation project: Using workstations as building blocks for parallel computing, to appear in Proc. of the International Conference on Parallel and Distributed Processing, Techniques and Applications (PDPTA96), New Horizons, Sunnyvale, CA, 09. - 11.08.1996.

2.   W. Schönauer, Scientific Computing on Vector Computers, North-Holland, Amsterdam, 1987.

3.   W. Schönauer, H. Häfner, A careful interpretation of simple kernel benchmarks yields the essential information about a parallel supercomputer, in Supercomputer 62 (1995), pp. 63 - 74.

Matrix Multiplication over Small Finite Fields on MIMD Architectures

Reiner Staszewski

Institut für Experimentelle Mathematik

Ellernstr. 29

D–45326 Essen

GERMANY

E-mail: reiner@exp-math.uni-essen.de

February 17, 1997

Summary

Many problems in computational representation theory of finite groups require calculations with huge dense matrices over very small finite fields (cf. [5] and [7]). In this paper, we concentrate on matrix multiplication of matrices over GF(2) though most of the ideas can be generalized in some way to other matrix algorithms and to other small finite fields.

In the first chapter, we present some ideas how to accelerate earlier programs and considerations how to choose the best possible corresponding tuning parameters. We show that these considerations and the running times of an implementation on one thin node of an IBM SP/2 fit well together.

In the second chapter, we discuss different approaches for the parallelization of the algorithms, which all use the message-passing paradigm. We study the influences between the tuning parameters and different communication patterns. This is explained looking at several examples performed on the SP/2.

1 Introduction

In this paper we want to consider the matrix operations over very small finite fields on a parallel computer with a distributed memory.

Our interest in this subject arose from computational modular representation theory of finite groups, where one would like to handle very large matrices over very small finite fields, but there are also other applications in algebra, coding theory and cryptography (cf. e. g. [4], [5] and [7]). The first programs which could deal with those matrices efficiently were written in Fortran by Richard Parker and Klaus Lux (cf. [3]), and they are around now for more than ten years. For a few years now, there is also a C version of Michael Ringe, which is distributed together with the Computer-Algebra System GAP (cf. [1]).

In this paper, we first present some ideas how to accelerate these programs and we afterwards discuss the influence of these ideas on the parallelization of the algorithms. We restrict ourselves to matrix multiplication over GF(2), though many ideas can be generalized to other small finite fields and to other row oriented algorithms and even to some column oriented ones (cf. [5] and [8]).

The efforts also had impact on the computational modular representation theory of finite groups. For example, using these parallel linear algebra programs on the SP/1 of the "Institut für Experimentelle Mathematik", Jens Rosenboom was able to prove in [5] that the last two unknown dimensions of the irreducible representations of the sporadic simple group Co_3 were 38,456 and 80,000 respectively. He did this by actually constructing them. The result was expected by Ibrahim A. I. Suleiman and Robert A. Wilson in [7] using probabilistic methods, but they were not able to prove this.

A lot of running time examples all performed on the SP/2 installation of the GMD, St. Augustin, Germany are given in the appendix.

2 The serial algorithms.

Throughout the whole paper, we'll stick to the following

Notation
Let $\mathcal{A} = (a_{ij})_{ij}$ be an $n \times m$-matrix, $\mathcal{B} = (b_{jk})_{jk}$ be an $m \times l$-matrix and $\mathcal{C} = (c_{ik})_{ik}$ their product $\mathcal{A} \cdot \mathcal{B}$, where all matrices are defined over GF(q), q a power of a prime.

Since we want to be able to handle non-sparse matrices with dimensions more than a hundred thousand, we are not allowed to waste any space, neither on the disk nor in the main memory. Thus we have to put several elements of our finite field $GF(q)$ into one machine word. If for example a machine word consists of four bytes, we pack 32 elements of $GF(2)$ into one word. Since in modular representation theory, matrices usually act from the right on the natural vector space by multiplication, the elements of our vector space are row vectors, so that the matrices are stored in row major order.

However extracting each b_{jk} of the kth column of B is now a very expensive operation in itself and extracting each a_{ij} to multiply it individually with each b_{jk} is even more expensive. Hence we cannot use the straightforward algorithm which calculates one c_{ik} after the other. Since the row $(c_{ik})_k$ of C is the sum of all rows $(b_{jk})_k$ each multiplied with the entry a_{ij} for $j = 1 \ldots m$, we use

Algorithm 1
Read A and put C to $\mathcal{O}$.
For j From 1 To m Do
   Read the jth row of B.
   For i From 1 To n Do
     If $a_{ij} \neq 0$ Then
       Add the a_{ij}-multiple of the jth row of B to the ith row of C.

This approach enables us to use efficient vector additions without the need to extract many single entries of the matrices, i. e. every entry of A is extracted once and no entry of B is extracted at all. This idea is used even in the earliest implementation by Richard Parker and others(cf. [3]).

Since we may expect our entries of A to be evenly distributed in the set of elements of $GF(q)$, the expected number of operations of this algorithm is $n \cdot m \cdot \frac{q-1}{q}$ row operations each of length l. If $q = 2$, then no field multiplication occurs and the addition of two matrix-rows is extremely simple. Supposing again that a machine word consists of four bytes, one exclusive–or operation on a machine word yields us the addition of 2 32-tuples of $GF(2)$. Under these circumstances, algorithm 1 essentially uses $\frac{nml}{64}$ integer-operations. But since the factor 32 is the same constant for all algorithms, we do not count the number of integer operations but the number of field operations.

As our field is finite and *small* and as there is no multiply-add instruction available for our field elements, it comes to mind to calculate all multiples of the just read in row of B, which may be reasonable if $q < n$ (and is always the case in our applications). Of course this is only sensible for $q > 2$. However, we can generalize this idea by using the concept of precomputing linear combinations of several rows of $\mathcal{B}$, so that it also makes a difference if $q = 2$. This leads to

Algorithm 2 *We fix an integer r.*
Read $\mathcal{A}$ and put $\mathcal{C}$ to $\mathcal{O}$.
For j From 1 To $\frac{m}{r}$ Do
    Read the next r rows of $\mathcal{B}$ and calculate all their linear combinations.
    For i From 1 To n Do
        If one of the next r columns of $\mathcal{A}$ Is Not zero Then
            Add the appropriate precalculated linear combination of the r rows of $\mathcal{B}$
            to the ith row of $\mathcal{C}$.

This idea has been around for quite some time and there are several implementations on different kind of machines using it. For example, there is a quite special serial implementation for the calculation of the semi-echelon-form of a matrix over GF(5) by Holger Gollan, an implementation of matrix multiplication and null space of a matrix over GF(5) on the CM-5, the last parallel machine of Thinking Machines Corporation by Holger Gollan, Michael Weller and the author, and an implementation of the null space of a matrix over GF(q) with $q \leq 256$ by Michael Weller, which uses the message passing concept on parallel machines (cf. [6] and [8]). All these implementations show that the idea is quite fruitful.

We now want to count the expected number of arithmetic operations for algorithm 2.

To calculate all linear combinations of r rows, we need $q^r - 1$ row-operations, where each row has length l; if q is a prime then every row-operation is even a row-addition. Hence the number of row-operations for the whole precomputation is

$$\frac{m}{r} \cdot (q^r - 1).$$

For the actual addition on $\mathcal{C}$ the number of row additions equals

$$n \cdot \frac{m}{r} \cdot \frac{q^r - 1}{q^r}.$$

since we as before expect the entries of $\mathcal{A}$ to be evenly distributed over GF(q).

Hence we have to find the natural number r such that the sum of the two expressions above gets minimal. The first function is monotone increasing (even exponentially) in r, the second is monotone decreasing in r, so that there exists a minimum also for natural numbers r, and this minimum is of course reasonably small. The value for r only depends on q and n, but not on m and l. So, for $q = 2$, the time t for algorithm 1 is just supposed to be $\frac{n}{2} \cdot m \cdot l$, whereas for algorithm 2, t should be

$$\frac{q^r - 1}{r} \cdot \left(1 + \frac{n}{q^r}\right) \cdot m \cdot l.$$

For $q = 2$ we get the following table:

1	$\leq n \leq$	4	$\Rightarrow r = 1$,	1.50 $m\,l$	$\leq t \leq$	3.00 $m\,l$
5	$\leq n \leq$	10	$\Rightarrow r = 2$,	3.38 $m\,l$	$\leq t \leq$	5.25 $m\,l$
11	$\leq n \leq$	24	$\Rightarrow r = 3$,	5.54 $m\,l$	$\leq t \leq$	9.33 $m\,l$
25	$\leq n \leq$	60	$\Rightarrow r = 4$,	9.61 $m\,l$	$\leq t \leq$	17.81 $m\,l$
61	$\leq n \leq$	144	$\Rightarrow r = 5$,	18.02 $m\,l$	$\leq t \leq$	34.10 $m\,l$
145	$\leq n \leq$	342	$\Rightarrow r = 6$,	34.29 $m\,l$	$\leq t \leq$	66.61 $m\,l$
343	$\leq n \leq$	797	$\Rightarrow r = 7$,	66.76 $m\,l$	$\leq t \leq$	131.11 $m\,l$
798	$\leq n \leq$	1828	$\Rightarrow r = 8$,	131.24 $m\,l$	$\leq t \leq$	259.48 $m\,l$
1829	$\leq n \leq$	4141	$\Rightarrow r = 9$,	259.60 $m\,l$	$\leq t \leq$	515.99 $m\,l$
4142	$\leq n \leq$	9271	$\Rightarrow r = 10$,	516.10 $m\,l$	$\leq t \leq$	1028.49 $m\,l$
9272	$\leq n \leq$	20546	$\Rightarrow r = 11$,	1028.59 $m\,l$	$\leq t \leq$	2053.00 $m\,l$
20547	$\leq n \leq$	45134	$\Rightarrow r = 12$,	2053.08 $m\,l$	$\leq t \leq$	4101.50 $m\,l$
45135	$\leq n \leq$	98395	$\Rightarrow r = 13$,	4101.58 $m\,l$	$\leq t \leq$	8198.00 $m\,l$

It for example tells us that algorithm 2 should be about 6 times faster than algorithm 1 if we deal with matrices of dimension 100,000. The practical behavior can be seen in Table 1. All our examples ran on the SP/2 installation of the GMD; this installation consists mainly of thin 66-MHz nodes, so that the serial runs were also made on this kind of node.

This table shows that even for very small matrices of dimension 1000, which only need 125,000 bytes of storage, we have a theoretical speedup of 3.2 and a practical speedup of 3.5 by choosing $r = 8$. The practical speedup is even bigger, because we didn't take into account the quadratic proportion which consists of looking at bits of $\mathcal{A}$ or at bytes of $\mathcal{A}$, respectively.

Comparing algorithm 2 to algorithm 1 we have decreased the number of the actually performed arithmetic operations. When dealing with matrices over floating-point numbers one reduces the number of loads and stores by using block-algorithms. If we handle a matrix over a small finite field, we can do something similar by reorganizing the j-loop such that several, say s, loops are handled simultaneously.

In C for example, it means that s loops of the form

for $(k = 0; k < lw; k + +)$
    $*ptr + +$ ^ $=$ $*ptr1 + +;$

are replaced by one loop of the form

for $(k = 0; k < lw; k + +)$
    $*ptr + +$ ^ $=$ $*ptr1 + +$ ^ $*ptr2 + +$ ^ $*ptr3 + +$ ^ $\cdots$ ^ $*ptrs + +;$

Here $ptr1, ptr2, ptr3, \cdots, ptrs$, and ptr are pointers to words, where $ptr1$, $ptr2, ptr3, \cdots, ptrs$ point to the beginning of the rows which we want to add and ptr points to the beginning of the row to which we want to add; moreover lw denotes the number of words in a single row of the matrix.

One step of the inner loop now consists of
$s + 1$ loads,
s exclusive-or operations,
1 store,
$s + 1$ pointer-incrementations.

For $s = 1$, this means for one step
2 loads,
1 exclusive-or operation,
1 store,
2 pointer-incrementations

and for s steps
$2s$ loads,
s exclusive-or operation,
s stores
$2s$ pointer-incrementations.

Since on most modern machines the memory bandwidth is the bottleneck for this kind of application, the new version should be more than twice as quick.

Hence we can improve the performance of algorithm 2 by decreasing the number of loads and stores and using

Algorithm 3 *We fix a further integer s.*
Read $\mathcal{A}$ and put $\mathcal{C}$ to $\mathcal{O}$.
For j From 1 To $\frac{m}{r \cdot s}$ Do
    For i From 1 To s Do

Read the next r rows of B and calculate all their linear combinations.

For i From 1 To n Do

Add the appropriate precalculated s different linear combinations of r rows of B to the ith row of C.

We want to analyze the effect of s in some more detail for an SP/2. The processors of the SP/2 can perform a pointer incrementation together with a load or a store not needing any extra time, so that we get actually

$s + 1$ loads,

s exclusive-or operations,

1 store

compared to

$2s$ loads,

s exclusive-or operation,

s stores.

On an SP/2, this gives a factor of much more than two. On a wide node for example, this factor should be about 3.6 for $s = 16$. This is due to the fact that the processor has two integer units, which can work in parallel, and two loads can be done in parallel, also. This can be used very efficiently for bigger s, but not for $s = 1$ or $s = 2$. Due to some loop unrolling of the compiler, the inner loop of the version with $s = 1$ takes 4.5 cycles, the version with $s = 2$ takes 5.5 cycles and the versions with $s \geq 3$ should take $s + 4$ cycles. This gives the factor of $\frac{4.5 \cdot 16}{20} = 3.6$ between $s = 1$ and $s = 16$. Unfortunately the C-Compiler is not able to produce such a code for $s = 15$ or $s = 16$, so that one has to do the writing in assembler to achieve such a performance. The Compiler is only able to do it in 23 cycles for $s = 16$ leading to a factor of 3.13.

Looking at Table 1 and Table 2 with $r = 1$, we see that the measured quotient on a thin node is $\frac{4811.70}{1399.42} = 3.44$. The same tables with $r = 8$ yield a factor of $\frac{1384.66}{544.82} = 2.54$. However, the calculation of the linear combinations is included in this time; subtracting this time of about 25 seconds gives $\frac{1359.45}{518.36} = 2.62$. This quotient is unfortunately much smaller than the earlier one; it's probably due to the fact that the number of cache misses increases if either r or s increase.

Looking at the timing for wide node, we get $\frac{4242.63}{1342.79} = 3.16$ for $r = 1$ and $\frac{1098.58 - 17.24}{375.92 - 20.11} = 3.04$ for $r = 8$. So these two quotients are much closer together, because the memory to cache bandwidth of a wide node is double this bandwidth of a thin node, so that the wide nodes are not so heavily effected by cache misses.

We want to make a few further remarks on Tables 2 and 3.

For the multiplication of two matrices of dimension 16,000, algorithm 3 gives us an acceleration of more than nine compared to algorithm 1.

The best possible values for r became less if $s = 16$. For example, if $n = 16,000$ the best possible value for r was 11, but it became 9 for $s = 16$. Hence the memory required for the linear combinations in this case only increased by a factor of four. This comes from the fact that s has no positive influence on the calculation of the linear combinations, because we always only have to add two rows together. Hence we have to choose our r such that the sum of the two expressions $\frac{q^r-1}{r}$ and $\alpha \cdot \frac{n}{r} \cdot \frac{q^r-1}{q^r}$ gets minimal, where α depends on s and the machine architecture. Getting the correct practical values r hence means to take the table on page 5 and multiply all the limits of the n's by $\frac{1}{\alpha}$. The running times on Tables 1 and 2 show that this approach gets good results in practice.

Moreover, performing the multiplication with $r = 12$ and $s = 16$ was impossible, because then the program required more than the available 128 megabytes of main memory. This shows that for bigger matrices r and s have to be chosen quite carefully, such that all still fits in memory.

3 The parallel algorithms.

We now want to discuss some different methods of parallelization, and we'll see the consequences of the choices of r and s for the data distribution. The first concept which comes to mind is a row-wise distribution of the data.

Suppose we have N processors available. Then we distribute the data in the following way and the matrix B is broadcasted to every processor.

$$
\mathcal{A} = \begin{pmatrix} \boxed{\begin{array}{c} P_0 \\ \hline P_1 \end{array}} \\ \cdot \\ \cdot \\ \cdot \\ \boxed{P_{N-1}} \end{pmatrix}
\qquad
\mathcal{C} = \begin{pmatrix} \boxed{\begin{array}{c} P_0 \\ \hline P_1 \end{array}} \\ \cdot \\ \cdot \\ \cdot \\ \boxed{P_{N-1}} \end{pmatrix}
$$

This approach has the advantage of not wasting any RAM.

If the Broadcast is done by a cyclic shift (cf. [8]), we obtain for algorithm 1 the running times of Table 3 and the efficiencies of Table 4, where we always look at the scaling for the same problem.

Compared with the results of H. Häfner and W. Schönauer about multiplication of matrices over floating-point numbers published in the same proceedings ([2]), we see at Table 3 that the efficiencies for matrices of dimension 8000 over GF(2) are a little bit less than for matrices of dimension 3000 over floating-point numbers. However, this is the efficiency for the slowest of the presented algorithms. The timings and efficiencies for algorithm 2 and algorithm 3 with $r = 8$ are given in Tables 5 to 10. One sees that corresponding efficiencies (for example the ones written in boldface) decrease from about 75% in Table 4 to 50% in Tables 6 and 8 and further to 40% and less in Table 10. This is to a large extend due to Amdahls law, because, with the row-wise distribution, the calculation of the linear combinations is a serial part of the program, since all processors have to calculate them completely.

Hence it naturally comes to mind to choose a column-wise distribution, because, in this case, every processor only has to calculate the linear combinations of those parts of the matrix B for which it is responsible. On the other hand, the communication pattern in this approach is much worse, because the master has to scatter each row of B to all the other processors. Hoping for good results in a compromise, we generalized both data distributions to an $N_x \times N_y$ mesh in the following way:

$$
\mathcal{C} \;=\; \begin{pmatrix}
\boxed{P_0} & \boxed{P_1} & \cdots & \boxed{P_{N_x-1}} \\
\boxed{P_{N_x}} & \boxed{P_{N_x+1}} & \cdots & \boxed{P_{2N_x-1}} \\
\cdot & \cdot & \cdots & \cdot \\
\cdot & \cdot & \cdots & \cdot \\
\cdot & \cdot & \cdots & \cdot \\
\boxed{P_{N-N_x}} & \boxed{P_{N-N_x+1}} & \cdots & \boxed{P_{N-1}}
\end{pmatrix}
$$

This means that the data distribution for the other matrices must be the following:

$$
\mathcal{A} \;=\; \begin{pmatrix}
\boxed{P_0, P_1, ..., P_{N_x-1}} \\
\boxed{P_{N_x}, P_{N_x+1}, ..., P_{2N_x-1}} \\
\cdot \\
\cdot \\
\cdot \\
\boxed{P_{N-N_x}, P_{N-N_x+1}, ..., P_{N-1}}
\end{pmatrix}
$$

$$
\mathcal{B} \;=\; \left(
\begin{array}{|c|c|c|c|}
\hline
\begin{array}{c} P_0 \\ P_{N_x} \\ \cdot \\ \cdot \\ \cdot \\ P_{N-N_x} \end{array} &
\begin{array}{c} P_1 \\ P_{N_x+1} \\ \cdot \\ \cdot \\ \cdot \\ P_{N-N_x+1} \end{array} &
\begin{array}{c} \cdots \\ \cdots \\ \cdot \\ \cdot \\ \cdot \\ \cdots \end{array} &
\begin{array}{c} P_{N_x-1} \\ P_{2N_x-1} \\ \cdot \\ \cdot \\ \cdot \\ P_{N-1} \end{array} \\
\hline
\end{array}
\right)
$$

Running times and efficiencies with $r = 8$ and $s = 16$ and this more general data distribution are given in Tables 11 and 12. The success of this data distribution is obvious; mostly a compromise similar to $N_x = N_y$ gives the best performance. Handling matrices of dimension 32,000 with 32 processors, the program with $N_x = 1$ and $N_y = 32$ needs about 40 % more time than the one with $N_x = 4$ and $N_y = 8$.

A detailed analysis of the communication costs compared with the calculation costs will be done in later paper. With this general distribution pattern the computing time and the communication time can be varied independently, so that these algorithms are very well suited for integer benchmarking.

The last two tables (13 and 14) present the practical "Best possible running times of algorithm 3" and the corresponding efficiencies, where s is always 16 (note that the matrices are still small and all the data fits into the main memory) and r is given also in Table 13. We see that we still have a loss of more than 30 % in the example mentioned above.

Moreover, we can see from the tables that the progress over the last few years of hard- and software together made to possible to accelerate matrix-multiplication over GF(2) by a factor of more than 125, even if the matrices are only of dimension 16,000, even though the efficiency is less than 40 %. Thus matrices of dimension 10 times as large as before can now be handled and Jens Rosenboom was able to do the work mentioned in the introduction (cf. [5]). Moreover, using the implementation of a null space described in [8] Peter Roelse (cf. [4]) was able to factor a polynomial of degree 300,000 over GF(2) in about $10\frac{1}{4}$ hours using 256 processors of the SP/2 installation of the Cornell Theory Center, USA.

Acknowledgments

I would like to thank the GMD, St. Augustin, Germany for the permission to use their SP/2 installation. Without access to their machine the experiments won't have been possible.

References

[1] M. Schönert et al. *GAP – Groups, Algorithms and Programming*, 1995.

[2] H. Häfner and W. Schönauer. Measurements, proper scaling and application of communication parameters. In G. Cooperman, G. Michler, and H. Vinck, editors, *Proceedings of the "Workshop on High Performance Computing and Gigabit Area Networks"*. Lecture Notes in Control and Information Sciences, Springer, 1997.

[3] R. A. Parker. The computer calculation of modular characters (the Meataxe). In M. D. Atkinson, editor, *Computational Group Theory*. Academic Press, London, 1984.

[4] P. Roelse. Factoring high-degree polynomials over F_2 with Niederreiter's algorithm on the IBM SP/2. In preparation, Institut für Experimentelle Mathematik, Ellernstraße 29, D 45326 Essen, Germany, 1997.

[5] J. Rosenboom. Verifying the 2-modular character table for Co_3. In G. Cooperman, G. Michler, and H. Vinck, editors, *Proceedings of the "Workshop on High Performance Computing and Gigabit Area Networks"*. Lecture Notes in Control and Information Sciences, Springer, 1997.

[6] R. Staszewski and M. Weller. Solving non-sparse systems of linear equations over finite fields on the CM-5. Preprint 21, Institut für Experimentelle Mathematik, Ellernstraße 29, D 45326 Essen, Germany, 1994.

[7] Ibrahim A. I. Suleiman and Robert A. Wilson. The 2-modular characters of conway's third group Co_3. *J. Symbolic Comput.*, To appear.

[8] M. Weller. Parallel gaussian elimination over small finite fields. In *Procedings of the 9th International Conference on Parallel and Distributed Computing Systems*. ISCA, September 1996.

Table 1: Running times of algorithm 2

$r \setminus n$	1000	2000	4000	8000	16000
1	1.84 s.	11.22 s.	81.94 s.	619.46 s.	4811.70 s.
2	1.28 s.	7.90 s.	58.58 s.	453.49 s.	3544.89 s.
3	0.99 s.	6.05 s.	44.98 s.	348.40 s.	2750.20 s.
4	0.96 s.	4.85 s.	35.89 s.	280.89 s.	2302.35 s.
5	0.69 s.	4.15 s.	30.12 s.	242.48 s.	2005.55 s.
6	0.60 s.	3.51 s.	26.61 s.	217.28 s.	1756.81 s.
7	0.55 s.	3.23 s.	24.68 s.	195.12 s.	1548.92 s.
8	**0.52** s.	**3.13** s.	22.84 s.	176.90 s.	1384.66 s.
9	0.57 s.	3.20 s.	**21.94** s.	163.95 s.	1262.95 s.
10	0.68 s.	3.45 s.	22.16 s.	**156.73** s.	1175.59 s.
11	0.86 s.	4.16 s.	23.96 s.	158.68 s.	**1136.56** s.
12	1.26 s.	5.65 s.	29.35 s.	174.73 s.	1151.51 s.

Table 2: Running times of algorithm 3 with $s = 16$.

$r \setminus n$	1000	2000	4000	8000	16000
1	0.91 s.	4.17 s.	26.84 s.	186.54 s.	1399.42 s.
2	0.52 s.	2.67 s.	17.19 s.	125.25 s.	1222.22 s.
3	0.44 s.	2.05 s.	13.69 s.	122.28 s.	1097.29 s.
4	0.36 s.	1.69 s.	13.29 s.	110.50 s.	930.31 s.
5	**0.33** s.	1.89 s.	12.55 s.	97.23 s.	787.48 s.
6	0.49 s.	1.76 s.	11.42 s.	85.42 s.	683.27 s.
7	0.37 s.	1.67 s.	10.68 s.	76.62 s.	604.90 s.
8	**0.38** s.	**1.65** s.	**9.91** s.	71.31 s.	544.82 s.
9	0.45 s.	1.82 s.	**10.56** s.	**69.72** s.	**518.32** s.
10	0.54 s.	2.26 s.	11.76 s.	**71.54** s.	521.87 s.
11	0.77 s.	3.13 s.	14.93 s.	81.17 s.	**593.06** s.
12	1.21 s.	4.76 s.	20.79 s.	102.66 s.	—— s.

Table 3: Running times of algorithm 1 on N nodes

$N \setminus n$	1000	2000	4000	8000	16000
1	1.84 s.	11.22 s.	81.94 s.	619.46 s.	4811.70 s.
2	**1.22** s.	6.47 s.	42.89 s.	318.28 s.	2418.82 s.
4	1.12 s.	**4.13** s.	23.35 s.	163.08 s.	1231.27 s.
8	1.19 s.	3.09 s.	**14.13** s.	86.99 s.	628.70 s.
16	2.06 s.	4.07 s.	11.24 s.	**50.75** s.	334.75 s.
32	3.14 s.	4.15 s.	10.10 s.	34.41 s.	**190.47** s.

Table 4: Efficiency of algorithm 1 on N nodes

$N \setminus n$	1000	2000	4000	8000	16000
1	100.00 %	100.00 %	100.00 %	100.00 %	100.00 %
2	**75.41** %	86.71 %	95.52 %	97.31 %	99.46 %
4	41.07 %	**67.92** %	87.73 %	94.96 %	97.70 %
8	19.33 %	45.39 %	**72.49** %	89.01 %	95.67 %
16	5.58 %	17.23 %	45.56 %	**76.29** %	89.84 %
32	1.81 %	8.45 %	25.35 %	56.26 %	**78.94** %

Table 5: Running times of algorithm 2 with $r = 8$

$N \setminus n$	2000	4000	8000	16000	32000
1	3.13 s.	22.84 s.	176.55 s.	1384.66 s.	11329.21 s.
2	1.97 s.	12.73 s.	93.75 s.	710.61 s.	5741.43 s.
4	**1.42** s.	7.90 s.	51.61 s.	376.02 s.	2925.72 s.
8	1.24 s.	**5.27** s.	31.28 s.	210.17 s.	1580.85 s.
16	2.08 s.	5.82 s.	**22.33** s.	126.54 s.	901.15 s.
32	2.25 s.	5.46 s.	17.89 s.	**87.98** s.	**557.54** s.

Table 6: Efficiency of algorithm 2 with $r = 8$

$N \setminus n$	2000	4000	8000	16000	32000
1	100.00 %	100.00 %	100.00 %	100.00 %	100.00 %
2	79.44 %	89.71 %	94.16 %	97.43 %	98.66 %
4	**55.11 %**	72.28 %	85.52 %	92.06 %	96.81 %
8	31.55 %	**54.17 %**	70.55 %	82.35 %	89.58 %
16	9.41 %	24.53 %	**49.42 %**	68.39 %	78.57 %
32	4.35 %	13.07 %	30.84 %	**49.18 %**	**63.50 %**

Table 7: Running times of algorithm 3 with $r = 1$ and $s = 16$

$N \setminus n$	1000	2000	4000	8000	16000
1	0.91 s.	4.17 s.	26.84 s.	186.54 s.	1399.42 s.
2	**0.87** s.	2.86 s.	14.85 s.	98.06 s.	711.75 s.
4	0.76 s.	**2.20** s.	9.19 s.	53.39 s.	372.79 s.
8	0.88 s.	2.10 s.	**6.70** s.	31.90 s.	199.42 s.
16	2.05 s.	3.64 s.	6.94 s.	**23.70** s.	118.57 s.
32	2.77 s.	3.87 s.	7.61 s.	19.88 s.	**82.13** s.

Table 8: Efficiency of algorithm 3 with $r = 1$ and $s = 16$

$N \setminus n$	1000	2000	4000	8000	16000
1	100.00 %	100.00 %	100.00 %	100.00 %	100.00 %
2	**52.30 %**	72.90 %	90.37 %	95.12 %	98.31 %
4	29.93 %	**47.39 %**	73.01 %	87.35 %	93.85 %
8	12.93 %	24.82 %	**50.07 %**	73.10 %	87.72 %
16	2.77 %	7.16 %	24.17 %	**49.19 %**	73.77 %
32	1.03 %	3.37 %	11.02 %	29.32 %	**53.25 %**

Table 9: Running times of algorithm 3 with $r = 8$ and $s = 16$

$N \setminus n$	2000	4000	8000	16000	32000
1	1.65 s.	9.91 s.	71.31 s.	544.82 s.	4861.47 s.
2	1.20 s.	6.31 s.	40.34 s.	291.36 s.	2462.20 s.
4	**1.06** s.	4.45 s.	24.97 s.	165.32 s.	1278.68 s.
8	1.10 s.	**3.72** s.	17.38 s.	102.14 s.	721.91 s.
16	2.09 s.	4.79 s.	**15.38** s.	73.05 s.	492.57 s.
32	2.78 s.	5.70 s.	17.01 s.	**61.88** s.	**379.78** s.

Table 10: Efficiency of algorithm 3 with $r = 8$ and $s = 16$

$N \setminus n$	2000	4000	8000	16000	32000
1	100.00 %	100.00 %	100.00 %	100.00 %	100.00 %
2	68.75 %	78.53 %	88.39 %	93.50 %	98.72 %
4	**38.92** %	55.67 %	71.40 %	82.39 %	95.05 %
8	18.75 %	**33.30** %	51.29 %	66.68 %	84.18 %
16	4.93 %	12.93 %	**28.98** %	46.61 %	61.69 %
32	1.85 %	5.43 %	13.10 %	**27.51** %	**40.00** %

Table 11: Running times of algorithm 3 with $r = 8$ and $s = 16$

	2000	4000	8000	16000	32000
1 × 1	1.65 s.	9.91 s.	71.31 s.	544.82 s.	4861.47 s.
2 × 1	1.20 s.	6.31 s.	40.34 s.	291.36 s.	2462.20 s.
1 × 2	1.25 s.	6.24 s.	39.35 s.	287.89 s.	2502.35 s.
4 × 1	**1.06 s.**	4.45 s.	24.97 s.	165.32 s.	1278.68 s.
2 × 2	**0.98 s.**	4.10 s.	23.26 s.	157.54 s.	1384.52 s.
1 × 4	**2.75 s.**	4.64 s.	24.42 s.	161.28 s.	1350.07 s.
8 × 1	1.10 s.	**3.72 s.**	17.38 s.	102.14 s.	721.91 s.
4 × 2	0.92 s.	**3.08 s.**	14.93 s.	89.65 s.	697.44 s.
2 × 4	0.97 s.	**3.10 s.**	13.95 s.	87.90 s.	696.94 s.
1 × 8	1.20 s.	**4.09 s.**	17.75 s.	97.55 s.	890.06 s.
16 × 1	2.09 s.	4.79 s.	**15.38 s.**	73.05 s.	492.57 s.
8 × 2	1.01 s.	2.85 s.	**11.01 s.**	58.81 s.	433.20 s.
4 × 4	2.11 s.	2.66 s.	**10.23 s.**	54.09 s.	409.51 s.
2 × 8	2.53 s.	3.20 s.	**13.02 s.**	57.99 s.	433.30 s.
1 × 16	2.47 s.	4.82 s.	**17.12 s.**	67.25 s.	538.14 s.
32 × 1	2.78 s.	5.70 s.	16.37 s.	**61.88 s.**	**379.78 s.**
16 × 2	2.43 s.	4.14 s.	10.86 s.	**43.79 s.**	**310.37 s.**
8 × 4	2.46 s.	2.95 s.	8.64 s.	**38.96 s.**	**273.89 s.**
4 × 8	2.47 s.	3.14 s.	8.88 s.	**37.68 s.**	**283.13 s.**
2 × 16	3.41 s.	5.25 s.	11.53 s.	**45.52 s.**	**294.99 s.**
1 × 32	3.91 s.	7.25 s.	17.05 s.	**61.00 s.**	**422.28 s.**

Table 12: Efficiency of algorithm 3 with $r = 8$ and $s = 16$

	2000	4000	8000	16000	32000
1 × 1	100.00 %	100.00 %	100.00 %	100.00 %	100.00 %
2 × 1	68.75 %	78.53 %	88.39 %	93.50 %	98.72 %
1 × 2	66.00 %	79.41 %	90.61 %	94.62 %	97.14 %
4 × 1	**38.92 %**	55.67 %	71.40 %	82.39 %	95.05 %
2 × 2	**42.09 %**	60.43 %	76.64 %	86.46 %	87.78 %
1 × 4	**15.00 %**	53.39 %	73.00 %	84.45 %	90.02 %
8 × 1	18.75 %	**33.30 %**	51.29 %	66.68 %	84.18 %
4 × 2	22.42 %	**40.22 %**	59.70 %	75.96 %	87.13 %
2 × 4	21.26 %	**39.96 %**	63.90 %	77.48 %	87.19 %
1 × 8	17.19 %	**30.29 %**	50.22 %	69.81 %	68.27 %
16 × 1	4.93 %	12.93 %	**28.98 %**	46.61 %	61.69 %
8 × 2	10.21 %	21.73 %	**40.48 %**	57.90 %	70.14 %
4 × 4	4.89 %	23.28 %	**43.57 %**	62.95 %	74.20 %
2 × 8	4.08 %	19.36 %	**34.23 %**	58.72 %	70.12 %
1 × 16	4.18 %	12.85 %	**26.03 %**	50.63 %	56.46 %
32 × 1	1.85 %	5.43 %	13.61 %	**27.51 %**	40.00 %
16 × 2	2.12 %	7.48 %	20.52 %	**38.88 %**	48.95 %
8 × 4	2.10 %	10.50 %	25.79 %	**43.70 %**	55.47 %
4 × 8	2.09 %	9.86 %	25.10 %	**45.18 %**	53.66 %
2 × 16	1.51 %	5.90 %	19.33 %	**37.40 %**	51.50 %
1 × 32	1.32 %	4.27 %	13.07 %	**27.91 %**	35.98 %

Table 13: Best possible running times of algorithm 3

		2000		4000		8000		16000		32000	
1 × 1		1.65	8	9.91	8	69.72	9	518.32	9	4767.21	9
2 × 1		1.15	4	6.31	8	40.34	8	282.88	9	2424.49	9
1 × 2		1.21	5	6.24	8	39.35	8	264.59	10	2484.66	9
4 × 1		**0.96**	4	4.23	7	24.97	8	165.32	8	1252.82	9
2 × 2		**0.54**	5	4.10	8	23.26	8	153.23	9	1208.59	9
1 × 4		**1.09**	6	4.64	8	24.42	8	154.26	9	1293.58	10
8 × 1		0.93	5	**3.27**	5	16.18	7	101.63	7	721.91	8
4 × 2		0.90	6	**3.00**	7	14.85	7	89.65	8	669.75	9
2 × 4		0.97	7	**3.10**	8	13.95	8	87.81	9	655.76	9
1 × 8		1.20	8	**4.01**	10	17.18	10	95.20	10	849.51	10
16 × 1		1.75	5	4.06	5	**13.22**	6	68.03	6	487.09	7
8 × 2		1.00	6	2.67	6	**10.74**	7	58.28	7	433.20	8
4 × 4		2.08	5	2.66	8	**10.23**	8	54.24	8	409.51	8
2 × 8		2.42	6	3.20	8	**11.61**	9	56.67	9	418.04	10
1 × 16		2.46	7	4.21	11	**14.53**	9	63.54	10	473.09	7
32 × 1		2.43	7	4.49	6	13.73	7	**52.49**	6	353.62	6
16 × 2		2.43	8	3.96	6	9.89	6	**41.96**	7	303.22	7
8 × 4		2.44	10	2.91	7	8.52	7	**38.26**	7	266.68	9
4 × 8		2.47	8	3.14	8	8.88	8	**37.68**	8	280.46	9
2 × 16		3.28	7	4.82	9	11.60	7	**43.87**	10	289.15	10
1 × 32		3.18	11	6.82	11	14.68	9	**52.68**	11	418.40	9

Table 14: Efficiency for the best possible running times of algorithm 3

			2000	4000	8000	16000	32000
1	×	1	100.00 %	100.00 %	100.00 %	100.00 %	100.00 %
2	×	1	71.74 %	78.53 %	86.42 %	91.61 %	98.31 %
1	×	2	68.18 %	79.41 %	88.59 %	97.95 %	95.93 %
4	×	1	**42.97 %**	58.57 %	69.80 %	78.38 %	95.13 %
2	×	2	**76.39 %**	60.43 %	74.94 %	84.57 %	98.61 %
1	×	4	**37.84 %**	53.39 %	71.38 %	84.00 %	92.13 %
8	×	1	22.18 %	**37.88 %**	53.86 %	63.75 %	82.55 %
4	×	2	22.92 %	**41.29 %**	58.69 %	72.27 %	88.97 %
2	×	4	21.26 %	**39.96 %**	62.47 %	73.78 %	90.87 %
1	×	8	17.19 %	**30.89 %**	50.73 %	68.06 %	70.15 %
16	×	1	5.89 %	15.26 %	**32.96 %**	47.62 %	61.17 %
8	×	2	10.31 %	23.20 %	**40.57 %**	55.59 %	68.78 %
4	×	4	4.96 %	23.28 %	**42.60 %**	59.73 %	72.76 %
2	×	8	4.26 %	19.36 %	**37.53 %**	57.16 %	71.27 %
1	×	16	4.19 %	14.71 %	**29.99 %**	50.98 %	62.98 %
32	×	1	2.12 %	6.90 %	15.87 %	**30.86 %**	42.13 %
16	×	2	2.12 %	7.82 %	22.03 %	**38.60 %**	49.13 %
8	×	4	2.11 %	10.64 %	25.57 %	**42.34 %**	55.86 %
4	×	8	2.09 %	9.86 %	24.54 %	**42.99 %**	53.12 %
2	×	16	1.57 %	6.43 %	18.78 %	**36.92 %**	51.52 %
1	×	32	1.62 %	4.54 %	14.84 %	**30.75 %**	35.61 %

Parallel Algorithms for Dense Eigenvalue Problems

Xiaobai Sun *

Department of Computer Science, Duke University, Durham, NC 27708 U.S.A.,
xiaobai@cs.duke.edu

Summary. Parallel solutions of dense eigenvalue problems have been active research topics since the implementation of the first parallel eigenvalue algorithm in 1971. In this paper we review Jacobi methods and spectral division methods in respective frameworks, introduce the issues arising for high performance implementation on contemporary computers, especially, parallel computers, and address problems that remain open.

1. Introduction

Numerical solutions of linear algebra problems, specifically, the solution of systems of linear equations, least squares problems, and algebraic eigenvalue problems, have been extensively used at kernels in computational solutions of large application problems in science and engineering. For eigenvalue problems, we are concerned with, in particular, the standard eigenvalue problem:

$$A\,x = \lambda x, \qquad x \neq 0, \tag{1.1}$$

and the general eigenvalue problem :

$$A\,x = \lambda B\,x, \qquad x \neq 0, \tag{1.2}$$

where A and B are square matrices of order n. To solve large problems efficiently on high-performance computers, especially, parallel computers, linear algebra algorithms are re-examined from the performance aspects such as efficiency, scalability and programmability as well as arithmetic complexity and numerical properties.

High-performance algorithms and parallel algorithms have been strongly influenced by the changes and varieties in architectures, compilers and operating systems. LAPACK, which stands for Linear Algebra PACKage [1], has successfully achieved high performance on a wide range of modern computers such as vector processors, super-scalar workstations and share-memory multiprocessors, due to substantial algorithm redesign and the specifications and implementations of BLAS [9]. The concepts and techniques of high performance computing recorded in LAPACK are of great value to the development

---

* This work was in part supported by the Advanced Research Projects Agency, under contract P-95006

of parallel algorithms. There are, however, more mathematical and computational issues involved in algorithm design for efficient parallel eigenvalue solvers on distributed-memory systems. Since the first parallel Jacobi algorithm implemented on ILLIAC IV in 1971 [25] the design and development of parallel algorithms for eigenvalue problems have been active research topics.

The solution of an eigenvalue problem of order n often involves dense matrix computations if a big portion of the eigenvalues and eigenvectors is required. The initial matrices are not necessarily dense. For example, a present approach to solving symmetric banded (non-tridiagonal) eigenvalue problems is to reduce the banded matrix into the tridiagonal form first. The matrix for the reduction transformations and hence for the eigenvectors is dense. Such banded eigengenvalue problems arise, for example, in quantum chemistry applications [12].

The algorithms that a great many efforts are given to implement fall into three classes: 1) methods via reduction of matrices to certain condensed forms, 2) Jacobi or Jacobi-like methods, and 3) spectral division methods. Many numerical and experimental advances have been made in the two-stage methods of class 1) for the symmetric eigenvalue problem (1.1), such as high-performance reductions to tridiagonal forms, see, for example, [11, 5, 15] and the divide-and-conquer solutions of the reduced tridiagonal eigenvalue problems [10, 14]. The difficulties of generalizing the approach to non-symmetric problems or general problems (1.2) turn many interests to methods of the other two classes. In this paper we present the Jacobi methods and spectral division methods in respective frameworks, discuss the issues arising for high performance implementation on contemporary computers, especially, parallel computers, and address problems that remain open. Since the methods of our interest can be described with arithmetic building blocks that can be easily implemented over a wide range of computers, we focus on the mathematical and numerical issues most of which arise in the pursuit for high performance. In Section 2., we describe Jacobi methods and their block variants, and we point out the considerable difference between two families of (parallel) block Jacobi methods. In Section 3. we introduce two dual families of spectral division methods, and presents a framework of spectral division methods based on what we call the CS-AB iterations. Concluding remarks are in Section 4..

2. Jacobi Methods

A symmetric matrix A has eigen-decompositions in the form of $A = Q\Lambda Q^{\mathrm{T}}$, where Q is an orthogonal matrix and Λ is diagonal matrix with the eigenvalues of A on the diagonal. A Jacobi method for symmetric problem (1.1) reduces the Frobenius norm of the off-diagonal portion of A with a sequence of orthogonal transformations so that the evolving matrix converges to a diagonal matrix and hence reveals all the eigenvalues. The Jacobi approach is the one of the earliest methods for symmetric eigenvalue problem, Jacobi's original paper appeared in 1846 in the literature [18]. The interest in Jacobi's approach is renewed due to its inherent parallelism. The first parallel algorithm, implemented in 1971, for the symmetric eigenvalue problem is based on the Jacobi method, see [25]. A recent study by Demmel and Veselić shows that Jacobi's method is more accurate than the QR method on symmetric positive definite matrices [8]. Moreover, the Jacobi approach can be extended to normal matrices or generalized to certain non-normal but structured matrices.

A (scalar) Jacobi methods uses plane rotations, also called Jacobi rotations, to reduce the norm of the off-diagonal portion. For a 2×2 symmetric eigenvalue problem, a Jacobi rotation J is an orthogonal matrix that diagonalizes the 2×2 symmetric matrix :

$$J^{\mathrm{T}} \left(\begin{array}{cc} a_{pp} & a_{pq} \\ a_{qp} & a_{qq} \end{array} \right) J. \tag{2.1}$$

There exist two such orthogonal matrices. A Jacobi rotation J of order greater than 2 is the same as the identity matrix except in rows and columns p and q. The submatrix of J at the intersection of rows and columns p and q is a 2×2 Jacobi rotation. For a given off-diagonal element $\alpha_{p,q}$, a Jacobi rotation, denoted by $J(p,q)$, can be determined to annihilate the element as of the 2×2 principal subblock shown in (2.1). It is important, however, to choose the rotation with the acute angle among the two candidates for convergence. The similarity transformation $J(p,q)^{\mathrm{T}} A J(p,q)$ alters the matrix A in only its rows and column p and q. A Jacobi method is characterized by the annihilating sequence, i.e., the sequence of index pairs $\{(p,q)|p \neq q\}$ to the off-diagonal elements to be eliminated successively. The sequence can be dynamic such as the classical Jacobi method, where each (p,q) is chosen so that $\alpha_{p,q}$ is the largest in modulus. Jacobi methods with static annihilating sequence are more suitable for high performance implementation. A static sequence is specified by its periodic segment, called a *sweep*, which covers all index pairs in $\{(p,q)|p < q \leq n\}$ without duplication. In the cyclic-by-row Jacobi method, for instance, a sweep annihilates every off-diagonal element once in the order $(1,2), (1,3), \ldots, (1,n); (2,3), (2,4), \ldots, (2,n); \ldots; (n-1,n)$. A sweep updates all the $O(n^2)$ elements of the evolving matrix and takes $O(n^3)$ floating-point operations(flops). The threshold Jacobi method is a sweep-based method with dynamic feature of skipping elements below sweep-dependent threshold [31]. The three particular algorithms mentioned here converge and converge asymptotically at quadratic rate.

In parallel computation, a sweep is reordered into a sequence of sets, each set consists of non-conflicting Jacobi rotations that solve up to $n/2$ independent 2×2 subproblems (2.1) and can be applied to the matrix A simultaneously. For the $n = 6$ case, for illustration, a sweep can be an ordering of the following 5 sets:

$$\begin{pmatrix} 1 & 2 \\ 3 & 4 \\ 5 & 6 \end{pmatrix}, \begin{pmatrix} 1 & 3 \\ 2 & 6 \\ 3 & 5 \end{pmatrix}, \begin{pmatrix} 1 & 6 \\ 4 & 5 \\ 2 & 3 \end{pmatrix}, \begin{pmatrix} 1 & 5 \\ 6 & 3 \\ 4 & 2 \end{pmatrix}, \begin{pmatrix} 1 & 3 \\ 5 & 2 \\ 6 & 4 \end{pmatrix}, \qquad (2.2)$$

where the index pairs of each set are given in rows. A sweep can thus be carried out in 5 parallel steps, each parallel step consists of 3 non-conflicting Jacobi rotations, see (2.2). The ordering of the parallel as listed in (2.2) is called Round-Robbin ordering for that it can be generated automatically from the first one by keeping the index 1 still and rotating the other $(n-1)$ indices a round counter-clockwise, see [25, 19, 20, 27] for more on parallel Jacobi orderings. The inherent data parallelism within a sweep and the easy parallel implementation make Jacobi methods natural candidates for parallel algorithms.

Jacobi methods are challenged in the following two aspects. First, according to the heuristic argument by Brent and Luk [6], the number of sweeps needed for convergence is proportional to $\log(n)$. In other words, the arithmetic cost of Jacobi methods is considered of $O(\log(n)n^3)$ flops, higher than other symmetric eigen-solvers. Secondly, the contemporary distributed-memory systems have high penalty on accesses to data on remote memories. Blocking scalar Jacobi methods seem an appropriate approach [13, 26, 3, 27, 16] to relieving both the problems.

Although Jacobi orderings on partitioned blocks of size $b > 1$ bear an analog to orderings over elements, there are considerable differences in the way the $2b \times 2b$ principal subblocks are dealt with in a algorithm. The parallel *blocked* algorithms proposed by Shroff and Shreiber [27] are essentially based on a class of elementwise orderings that allow being reordered into two-level sub-sweeps: one block-wise sweep over off-diagonal blocks and one element-wise sweep over 'non-reserved' off-diagonal elements within $2b \times 2b$ principal blocks. The reserved elements are those off-diagonal elements in diagonal blocks and are to be annihilated only once in a sweep. The convergence analysis for elementwise orderings applies, and the number of sweeps remains therefore the same. The extra in arithmetic cost is mainly from the overhead in aggregating rotations in each block.

Many recent implementations of parallel Jacobi algorithms are based on *block* Jacobi methods instead, perhaps due to the straightforward extension from their scalar counterparts. Also, it is hoped that such blocking can reduce the number of sweeps. A block Jacobi method diagonalizes each $2b \times 2b$ principal block in a sweep over off-diagonal blocks [13] as scalar Jacobi methods do on 2×2 blocks (2.1). Note that Shroff-Schreiber methods do not diagonalize the $2b \times 2b$ principal blocks. As for the eigenvalue problem itself, there are many candidates for the diagonalization of subblocks. Some of the block algorithms may cause high tradeoff of arithmetic complexity. The following table presents a few snapshots of a comparison between two block-row-cyclic Jacobi methods, one using QR algorithm for the block diagonalization, the other using the scalar-row-cyclic Jacobi algorithm. The algorithms are implemented in MATLAB, and the matrix entries are generated from a uniform distribution on the interval (0.0,1.0).

<u>Table 1.</u>

$n = 30$	QR within blocks	Jacobi within blocks
block size	sweep numbers	
1	7	7
2	27	7
3	29	7
5	21	6
6	17	6
10	6	5

Notice that the block-row-cyclic method with Jacobi within blocks becomes the row-cyclic Jacobi at both block size 1 and block size 15. If a block algorithm with block size $b > 1$ takes as many sweeps as a non-block algorithm, then the arithmetic complexity is increased by a same ratio of a block diagonalization to an elementwise sweep over the block. We have observed that the number of sweeps is sensitive to the block diagonalization procedure and the block size. Using the QR algorithm within blocks, the block-row-cyclic method with smaller block size takes many more sweeps than the nonblock algorithm. The extra sweeps downgrade significantly the effect of blocking for high performance. The selection in diagonalization procedure is critical to the convergence as the selection of the 2-dimensional rotation with the acute angle. The difficulty in selecting a diagonalization procedure for convergence and faster convergence rate lies in the fact that the diagonalization procedures are in general iterative. Recently, Mackey [21] has proposed the so called quaternion-Jacobi methods, using 4-dimensional rotations to diagonalizes 4×4 subblocks, and proved the convergence. Mackey observed that quaternion-Jacobi method reaches the range of the quadratic convergence sooner than the plane-Jacobi method. We are still lack of general convergence analysis and numerical analysis for block Jacobi methods with two-levels or multi-levels iterations.

3. Spectral Division Methods

The basic computational step for a spectral division method is to compute two complementary subspaces according to a division in the spectrum. In particular, for problem (1.1), we compute an orthogonal matrix $Q = (Q_1, Q_2)$ such that

$$Q^{\mathrm{T}} A Q = \begin{pmatrix} A_{11} & A_{12} \\ 0 & A_{22} \end{pmatrix}, \tag{3.1}$$

where $\lambda(A_{11})$, the set of the eigenvalues of A_{11}, is disjoint from $\lambda(A_{22})$, Q_1 represents the right invariant subspace corresponding to $\lambda(A_{11})$, and Q_2^{H}

represents the left invariant subspace corresponding to $\lambda(A_{22})$. The problem is either solved if Q_1 or Q_2 is the solution of interest, or is reduced to smaller ones and the same idea can be applied, recursively, to the reduced problems. There are two kinds of spectral division methods known in the literature: one is based on the Newton iteration for the matrix sign function; another is based on Malyshev's iteration for what we shall call the matrix *disc* function. In this section, we describe the basic methods and then present a framework of inverse-free spectral division methods.

The common approach used in spectral division methods is to first compute a matrix function $F(A)$ that maps a specified portion of the eigenvalues of A to zero and then separate the rank-space and nullspace of $F(A)$ with rank revealing techniques to get Q. The chosen matrix function should be easily and efficiently implemented. For example, the shifted matrix sign function $\text{sign}(A) + I$ maps all eigenvalues in the left complex plan to zero, where I is the identity matrix. A well-known procedure to compute $\text{sign}(A)$ is the Newton iteration

$$A_{k+1} = (A_k + A_k^{-1})/2, \quad k \geq 0, \tag{3.2}$$

where $A_0 = A$ is nonsingular. Due to the inverse operation at each step, the Newton iteration is sensitive to perturbations and rounding errors if matrix A is poorly conditioned. In recent years, the work of Bulgakov, Godunov and Malyshev, see [22], and of Bai, Demmel and Gu [2], has brought into light the methods using the characteristic set function that maps the eigenvalues $|\lambda| < 1$ to 1 and the eigenvalues $|\lambda| > 1$ to 0. We may call the function *disc*. The disc function and the shifted sign function are dual to each other in the sense that the subsets divided by one can be mapped to those divided by the other under Cayley transform. The sequence $(1 + \lambda^k)^{-1}$, $|\lambda| \neq 1$, converges to $\text{disc}(\lambda)$. The following algorithm, based on Malyshev's work, computes $\text{disk}(A)$.

Algorithm 1. Computation of $\text{disc}(A)$

(Assumption: no eigenvalues on the unit circle)
1) $A_0 = A$; $B_0 = I$.
2) Malysheve iteration.
  For $j = 0 : 1 : 2, \ldots$, until converge

$$\begin{pmatrix} U_{11} & U_{12} \\ U_{21} & U_{22} \end{pmatrix} \begin{pmatrix} B_j \\ -A_j \end{pmatrix} = \begin{pmatrix} R_j \\ 0 \end{pmatrix}, \quad \text{QR factorization}$$

$$A_{j+1} = U_{21}A_j; \quad B_{j+1} = U_{22}B_j,$$

3) $\text{disc}(A) = (A_\infty + B_\infty)^{-1}B_\infty$.

Notice that step 2) involves no inverse operation, the basic operations are QR factorizations and matrix-matrix multiplications, which can be implemented efficiently on a wide range of parallel computers. Bai, Demmel and Gu [2] combine step 3), which re-introduces inverse operation, and the following step of separating the nullspace and range of disc(A) with generalized rank revealing technique to make Algorithm 1 inverse free. Sun and Quintana–Ortí [30] suggest an alternative way to compute the invariant subspaces without resorting to disc(A) and hence eliminating the inverse operation. In the following we describe from a simpler and broader perspective a family of inverse free spectral methods.

3.1 The CS-AB Iteration

Algorithm 1 transforms a standard eigenvalue problem (1.1) to a general problem (1.2) by setting the initial matrix pencil $(A_0, B_0) = (A, I)$,

$$A_0 x = \lambda B_0 x, \qquad |\lambda| \neq 1. \tag{3.3}$$

For $k \geq 0$, let (S_{k+1}, C_{k+1}) be a matrix pencil satisfying the *CS-AB equivalence rule*:

$$C_{k+1} A_k = S_{k+1} B_k, \tag{3.4}$$

we define the *the CS-AB recursion rule*:

$$A_{k+1} = S_{k+1} A_k, \quad B_{k+1} = C_{k+1} B_k. \tag{3.5}$$

Notice that the matrix pencil (U_{21}, U_{22}) in Algorithm 1 is a special choice for (S_{k+1}, C_{k+1}) obtained by a QR factorization. For $|\lambda| < 1$, we have

$$A_{k+1} x = S_{k+1} A_k = \lambda^{2^k} S_{k+1} B_k = \lambda^{2^k} C_{k+1} A_k x = \lambda^{2^{k+1}} B_{k+1} x, \tag{3.6}$$

and for $|\lambda| > 1$, we let $\nu = \lambda^{-1}$ and have

$$B_{k+1} x = C_{k+1} B_k x = \nu^{2^k} C_{k+1} A_k x = \nu^{2^k} S_{k+1} B_k x = \nu^{2^{k+1}} A_{k+1} x. \tag{3.7}$$

Consequently,

$$\begin{aligned}
(A_k + B_k) x = (1 + \lambda^{2^k}) B_k x, & \quad |\lambda| < 1, \\
(A_k + B_k) x = (1 + \nu^{2^k}) A_k x, & \quad |\nu| < 1.
\end{aligned} \tag{3.8}$$

Note that $\{1 + \lambda^{2^k}\}_{k=0}^{\infty}$ is a subsequence of $\{1 + \lambda^k\}_{k=0}^{\infty}$.

We can draw the following conclusions about the CS-AB iterations: 1) the pencil sequence (A_k, B_k) preserves the right eigenvectors or invariant subspaces and successively squares the eigenvalues, 2) if (A_k, B_k) converges, say, to (A_∞, B_∞), then the null spaces of A_∞, denoted by null(A_∞), and null(B_∞) contain, respectively, the invariant subspaces X_{in} and X_{out} corresponding to the eigenvalues of A inside and outside the unit circle. We can prove that if (A_0, B_0) is regular and has no eigenvalue at the unit circle, then null(A_∞) $\cap$ null(B_∞) = $\{0\}$, which implies that $(A_\infty + B_\infty)$ is nonsingular. From (3.8) and the equality $(A_\infty + B_\infty)^{-1}(A_\infty + B_\infty) = I$, we have

$$(A_\infty + B_\infty)^{-1} B_\infty (X_{in}, X_{out}) = (X_{in}, 0),$$
$$(A_\infty + B_\infty)^{-1} A_\infty (X_{in}, X_{out}) = (0, X_{out}).$$

In other words, $(A_\infty + B_\infty)^{-1} B_\infty = \text{disc}(A)$ and $(A_\infty + B_\infty)^{-1} A_\infty = I - \text{disc}(A)$. We observe that one can simply exploit the fact that $B_\infty X_{out} = 0$ and extract the invariant subspace X_{in} using rank-revealing techniques [30].

3.2 The convergence conditions

Recall that the QR factorization in the iteration step 2) of Algorithm 1 generates matrix pencils satisfying the CS-AB equivalence rule (3.4). There are other ways to generate (C_k, S_k) satisfying the CS-AB equivalence rule. For instance, an LR factorization can be used to substitute for the QR factorization in step 2) of Algorithm 1. Such an iteration may not converge. For instance, a plain use of the LR factorization may result in $B_k = I$,

$$\begin{pmatrix} I & 0 \\ S_{k+1} & I \end{pmatrix} \begin{pmatrix} I \\ -A_k \end{pmatrix} = 0,$$

$C_{k+1} = I$ and $S_{k+1} = A_k$, for all $k \geq 0$. Thus, $A_{k+1} = A_k^2 = A^{2^k}$, i.e., A_k is squared explicitly and hence does not converge as $k \to \infty$(we assume that A has eigenvalues on both sides of the unit circle.) Although $(A_k + B_k)^{-1} B_k$ converges to $\text{disc}(A)$, it is not a computationally appropriate procedure. To be computationally feasible, it is necessary to make the elements of (A_k, B_k) bounded. The QR factorization in Algorithm 1 guarantees that $(\|A_{k+1}\|_2 + \|B_{k+1}\|_2) \leq \|A_K\|_2 + \|B_k\|_2$.

We have mentioned in an earlier discussion that if (A_k, B_k) converges, then both A_∞ and B_∞ are singular. This necessary condition indicates that (A_k, B_k) generated by a CS-AB iteration can not be convergent if either A_k or B_k is kept nonsingular or even orthogonal. If an LR factorization is to be employed, the use of a non-uniform scaling or weighting strategy on both A_k and B_k is essential.

The convergence of spectral division methods also depends on the eigenvalue distribution. There is no spectral division if all the eigenvalues are on one side of the dividing boundary, the imaginary axis for the Newton iteration (3.2) or the unit circle for CS-AB iterations. If there are eigenvalues on the dividing boundary, spectral methods fail to converge.

For symmetric matrices and skew symmetric matrices, there are simplified and stabilized spectral methods without breakdown, even in the presence of eigenvalues at the dividing boundary [29].

3.3 Computation costs

The Newton iteration for matrix sign function and the Malyshev iteration for the matrix disc function are dual to each other as well under Cayley transformation. Both approaches therefore converge at the same pace. To be practically competitive, inverse free spectral methods need to decrease arithmetic cost in two aspects. First, the Newton iteration costs per step much less than the Malyshev iteration. We can show that a carefully designed QR factorization procedure can keep U_{22} and B_k triangular and hence reduces significantly the cost of per Malyshev iteration significantly. The requirement for space is reduced as well. The implementation is in progress. Secondly, there are effective acceleration techniques for the Newton iteration. An interesting research topic is to find dual acceleration techniques for the Malyshev iteration.

4. Concluding Remarks

Both Jacobi methods and spectral division methods have high potential for parallelism, and yet need to be better understood and further explored. Jacobi methods are more suitable for normal eigenvalue problems and similarly structured eigenvalue problems. Spectral division methods are often preferred in control application problems where an invariant subspace corresponding to a specific set of eigenvalues is of interest. A key component of spectral division methods, with or without inverse, is the separation of the range and the nullspace of a matrix. Some effective numerical algorithms for the space separation have been developed in recent years, such as rank-revealing QR factorizations (RR-QRF) [4, 7, 17, 23, 24], and rank-revealing URV [28]. Many spectral divisions other than along the imaginary axis or the unit circle can be achieved by choosing linear combinations of A and B for the initial matrix pencil [2, 30]. By nature, inverse free spectral division methods are methods for the generalized eigenvalue problem. It is shown in [30] that the cost of inverse free methods for generalized Schur decomposition can be reduced to about the same as for standard Schur decomposition.

Reference

1. E. Anderson and et al. *LAPACK User's Guide*. SIAM, Philadelphia, 2nd edition, 1995.
2. Z. Bai, J. W. Demmel, and M. Gu. Inverse free parallel spectral divide and conquer algorithms for nonsymmetric eigenproblems. Research Report 94-01, Department of Mathematics, University of Kentucky, 1994.
3. C. H. Bischof. The two-sided block Jacobi method on hypercube architectures. In M. T. Heath, editor, *Hypercube Multiprocessors*. SIAM Publications, Philadelphia, 1987.
4. C. H. Bischof and G. Quintana-Orti. Computing rank-revealing QR factorizations of dense matrices. *to appear on ACM: Trans. of Math. Soft.*, 1996.

5. C. H. Bischof and X. Sun. A framework for band reduction and tridiagonalization of symmetric matrices. T.R. MCS-P298-0392, Argonne National Laboratory, Mathematics and Computer Science Division, 1992.

6. R. P. Brent and F. T. Luk. The solution of singular-value and symmetric eigenvalue problems on multiprocessor arrays. *SIAM Journal on Scientific and Statistical Computing*, 6:69–84, 1985.

7. T. Chan. Rank-revealing QR factorizations. *Lin. Alg. Appl.*, 88/89:67–82, 1987.

8. J. Demmel and K. Veselić. Jacobi's method is more accurate than QR. *SIAM J. Matrix Anal. Appl.*, 13:1204–1245, 1992.

9. J. J. Dongara, J. DuCros, I. Duff, and S. Hammarling. A set of level 3 Basic Linear Algebra Subprograms. *ACM Trans. Math. Softw.*, 16:1–17, 1990.

10. J. J. Dongarra and D. C. Sorensen. A fully parallel algorithm for the symmetric eigenvalue problem. *SIAM Journal on Scientific and Statistical Computing*, 8:s139–s154, 1987.

11. J. J. Dongarra and R. A. van de Geijn. Reduction to condensed form for the eigenvalue problem on distributed memory architectures. Report ORNL/TM-12006, Mathematical Sciences Section, Oak Ridge National Laboratory, 1991.

12. G. Fan, R. J. Littlefield, and D. Elwood. Performance of a fully parallel dense real symmetric eigensolver in quantum chemistry applications. In *Proceedings of High Performance Computing '95*, Phoenix, AZ, Apr. 1995. Society for Computer Simulation.

13. G. H. Golub and C. F. Van Loan. *Matrix Computations*. The Johns Hopkins University Press, 2nd edition, 1989.

14. M. Gu and S. Eisenstat. A divide-and-conquer algorithm for the symmetric tridiagonal eigenproblem. *SIAM J. Matrix Anal. Appl.*, 16:171–191, 1995.

15. B. Henderckson, E. Jessup, and C. Smith. A parallel eigensolver for dense symmetric matrices. Technical report, Sandia National Labs, Univ. of Colorado, and Advanced Product and Process Tech., March 1996.

16. Vicente Hernandez, Robert A. van de Geijn, and Antonio M. Vidal. A Jacobi method by blocks on a mesh of processors. *Concurrency: Practice and Experience*, 1996. to appear.

17. P. Hong and C. T. Pan. The rank revealing QR and SVD. *Math. Comp.*, 58:213–232, 1992.

18. C. G. J. Jacobi. Über ein leichtes Verfahren die in der Theorie der Säculärstörungen vorkommenden Gleichungen numerisch aufzulösen. *Journal für die reine und angewandte Mathematik*, 30:51–s94, 1846.

19. F. T. Luk. Computing the singular-value decomposition on the ILLIAC IV. *ACM Transactions on Mathematical Software*, 6:524–539, 1983.

20. F. T. Luk and H. Park. On parallel jacobi orderings. *SIAM Journal on Scientific and Statistical Computing*, 10:18–26, 1989.

21. Niloufer Mackey. Hamilton and Jacobi meet again: quaternions and the eigenvalue problem. *SIAM J. Matrix Annal. Appl.*, 16:421–435, Apr. 1995.

22. A. N. Malyshev. Parallel algorithm for solving some spectral problems of linear algebra. *Lin. Alg. Appl.*, 188,189:489–520, 1993.

23. G. Quintana, X. Sun, and C. H. Bischof. A BLAS-3 version of the QR factorization with column pivoting. Technical Report MCS-P551-1295, Mathematics and Computer Science Division, Argonne National Laboratory, 1996. *To appear in SIAM J. Sci. Comp.*

24. G. Quintana-Ortí and E. S. Quintana-Ortí. Parallel bidimensional algorithms for computing rank-revealing QR factorizations. In *in the same proceeding*, 1996.

25. A. Sameh. On Jacobi and Jacobi-like algorithms for a parallel computer. *Mathematics of Computation*, 25:579–590, 1971.

26. D. S. Scott, M. T. Heath, and R. C. Ward. Parallel block Jacobi eigenvalue algorithms using systolic arrays. *Linear Algebra and Its Applications*, 77:345–355, 1986.

27. G. Shroff and R. Schreiber. On the convergence of the cyclic Jacobi method for parallel block ordering. *SIAM J. on Matrix Anal. Appl.*, 10(3):326–346, 1989.

28. G. W. Stewart. An updating algorithm for subspace tracking. *IEEE Trans. Signal Processing*, 40:1535–1541, 1992.

29. X. Sun. Scaling and squaring in invariant subspace decomposition methods. In *Large-Scale Matrix Diagonalization Methods*. Argonne national Laboratory, Chemistry Theory Institute, May 1996.

30. X. Sun and E. S. Quintana-Ortí. Spectral division methods for block generalized Schur decompositions. Technical Report CS-1996-13, Duke University,Department of Computer Science, August 1996.

31. J. H. Wilkinson. *The Algebraic Eigenvalue Problem*. Claredon Press, Oxford, England, 1965.

Enhanced Communication Support for Networked Applications

Martina Zitterbart
Institute of Operating Systems and Computer Networks
TU Braunschweig
zit@ibr.cs.tu-bs.de
http://www.ibr.cs.tu-bs.de

Abstract

Today communication support forms a key issue for almost all computer-driven applications. Enhanced networks as well as communication protocols and services are required in order to appropriately serve the emerging variety of applications. This holds for networked multimedia applications as well as for scientific applications. If scientific applications are implemented on workstation clusters, they typically need to deal with the same protocol stack as multimedia applications, i.e., with the Internet protocols. Although both types of applications may have different flavours, many commonalities exist with respect to their communication requirements. Examples include low communication overhead as well as scalable and efficient support of group communication.

The paper gives a survey of enhanced communication support that is mainly targeted towards services and protocols for networked multimedia applications. Since efficiency of communication systems is very important, implementation techniques are also considered within this survey.

1. Introduction

The spectrum of networked applications is becoming very wide, including multimedia applications as well as scientific applications (e.g., weather prediction, crash simulation and the like). The variety of networks is also increasing. Implementing scientific applications on workstation clusters instead of supercomputers is becoming more and more popular due to cost efficiency reasons. Networking and communication support is inherently needed for all these types of applications. Concrete communication requirements, however, are usually highly application dependent.

1.1 Application Requirements

Typical examples of multimedia applications include teleconferencing and CSCW. They comprise various components, such as audio, video, data, and shared data (e.g., whiteboard). Networked multimedia applications can be characterized by high and diverse requirements on the underlying communication system, specifically including
- support for group communication,
- low latency,
- continuous data delivery and, possibly,
- high data volume.

Compared to that, scientific applications typically require low communication overhead and fully reliable data transfer. Group communication facilities are also of interest. Continuous data streams are usually not needed.

The paradigm of group communication is inherent for many multimedia applications. Thus, it should be efficiently supported by all involved components including networks, protocols and the computer architecture itself. Efficiency and scalability of group communication currently form very important and vital research issues. In this context, scalability does not only address the size of the group, but also aspects like geographical distribution and heterogeneity of group members (e.g., different video or processing capabilities).

In terms of the OSI reference model, group communication support is not limited to a single layer. The network layer must support multicast addressing and routing. Error control in the transport layer must be suited for group communication. Furthermore, group management forms an integral part. It is commonly located somewhere in the middleware [Bern96] and may even include application specific subtasks (e.g., in the ITU-T T.120 standards [T120]). The emerging Internet protocols discussed in section 3 address some of these issues. The services that are currently provided by the Internet protocols are targeted towards traditional data applications (e.g., file transfer). Multimedia applications, however, require new services, partly with quality-of-service (QoS) guarantees.

1.2 Network Capabilities

High performance networks with integrated services are under development, most prominently represented by ATM-networks [Alle95]. Switched local area networks like Fast Ethernet and Switched Token Ring also provide higher data rates than traditional networks and include additional functionalities that are mostly directed towards multimedia support. For example, flow control mechanisms can be introduced in order to better support multimedia applications over Fast Ethernet [John96].

According to the ATM Forum [ATMF95], the following services are provided by ATM-based networks:

- CBR: Constant Bit Rate,
- rt-VBR: Real-Time Variable Bit Rate,
- nrt-VBR: Non-Real-Time Variable Bit Rate,
- UBR: Unspecified Bit Rate, and
- ABR: Available Bit Rate.

ATM-services are basically distinguished into real-time services (CBR, rt-VBR) and non-real-time services (nrt-VBR, UBR and ABR). Real-time services are designed for the support of audio and video applications that are sending data with constant or variable bit rate. Examples for nrt-VBR services include time critical transaction processing as, for example, needed for bank transactions. UBR services are meant for the support of traditional data applications like file transfer. For UBR services, no guarantees are given. ABR services provide end-to-end service guarantees at the network level. Scientific applications can be categorized as typical users of ABR services.

Emerging ATM-based networks implement service integration. Thus, ATM networks provide a sound basis for the implementation of networked multimedia applications. However, there are still issues to be solved for ATM networks, e.g., considering scalable and efficient support of group communication.

Generally, high performance networks are not sufficient in order to provide enhanced communication support for networked applications. Attention to higher layer protocols as well as to the system architecture and to implementation techniques is needed. The latter also contribute considerably to the quality of service that can be experienced by the user.

1.3 Outline

The paper is structured as follows. Section 2 concentrates on services needed for the support of networked application with special focus on multimedia applications. Emerging protocols in the Internet that are suited as the basis for multimedia applications are discussed in section 3. Implementation techniques that are critical for efficient communication systems are subject of section 4. Section 5 summarizes the paper.

2. Services

Within this section, communication services are discussed. The main focus is on distributed multimedia applications. Different application categories as well as different levels of service support within communication systems are presented.

2.1 Application Categories

In order to determine the service requirements of networked multimedia applications, they can be categorized in classes with potentially different requirements (cf., Table 1):

Table 1 Application Categories

Data Flow	Style	Timing Requirements	Typical User Requirements
asym-metric	broad-casting	isochronous data	low setup latency (few seconds) high downstream bandwidth (several Mbps) TV or HIFI quality (colour, resolution, ...)
	playback	isochronous data (short/long streams) interactivity	high downstream bandwidth (video clips) low upstream bandwidth (user interactions) low upstream and downstream latency TV or HIFI quality (colour, resolution, ...)
	real-time	interactivity	typically low bandwidth and latency guaranteed maximum reaction time
symmetric	conferencing	isochronous data interactivity	high bandwidth requirements very low latency high video/audio quality (video size, resolution) synchronization (mostly audio and video)
	collaboration	interactivity isochronous data	moderate bandwidth requirements very low latency synchronization (very fine granularity)

- *broadcasting applications*: typical examples are Digital Video Broadcast (DVB) and Digital Audio Broadcast (DAB).
- *interactive playback*: typical examples include Video-on-Demand (VCR-like) and teaching/tutoring applications. Mostly, playback is intended for a single individual user (or for a single end system, e.g., TV).
- *conferencing*: conferencing applications include audio and video streams. Their main focus is on these two media. Data applications, e.g., whiteboard applications, may be incorporated to support conferencing.

- *collaboration:* we classify applications as collaborative, that focus on the collaborative data part. Audio and video may just be used as supportive tools. Typical examples of collaborative applications are joint editing and interactive games.
- *traditional real-time applications:* A typical example of real-time applications can be seen in tele-robotics. Furthermore, many scientific applications may also be categorized as real-time applications. They need to receive the results of multiple processors within a bounded time interval.

Two basic types of timing requirements are distinguished:

- requirements from isochronous data streams, and
- requirements due to interactivity.

Isochronous requirements are introduced by applications that include audio and/or video streams which require continous data delivery to the user. Therefore, guaranteed upper bounds on delay jitter are needed. The requested QoS needs to be supported without any disruption during the entire communication phase. Reliability requirements of isochronous applications differ from those of traditional data applications:

- reliability is tied to timelines (i.e., data received after a deadline is considered as lost),
- no absolute reliability is needed (error tolerance depends on coding and compression).

Interactivity basically requires very low response times to user actions and, thus, very low latency. High bandwidth is not necessarily involved. Traditional file transfer applications have no explicit timing requirements. Within the Internet Services framework [BrCS94] they are, therefore, named elastic applications. Timing requirements may, however, be more strict if they are part of an application which consists of multiple streams (audio, video, data) that need to be synchronized. This implicitly may lead to high service requirements on such traditional applications as well. Generally, completely independent streams may have different requirements than streams being part of a more complex application.

Moreover, many forthcoming applications are inherently based on *multicasting*. This should be seen as a major communication paradigm for the future and, thus, should also be directly reflected in any attempt to support different types of services. For example, it leads to heterogeneous QoS support within a group due to different network attachment characteristics.

All components involved in the communication need to address these QoS requirements, including operating systems and system architectures (e.g., bus access latency and jitter). Generally, in order to support multimedia applications, user-to-user services - as they are finally required - need to be addressed. Therefore, different QoS levels, as discussed below, need to be distinguished.

2.2 QoS Levels

Since QoS is related to the dedication of shared resources to networked applications, it needs to be considered at different levels and for various types of resources. Three QoS levels are sufficient (cf., Figure 1):
- user level,
- application level, and
- network level.

The *user level* is directly related to human perception issues. Typically, users state their QoS requirements in an imprecise and highly subjective manner. Examples are type and size of a video window (e.g., colour and large) in case of videoconferencing. However, the parameter large does not reflect a precise value since it is relative to the size of the screen used. Additionally, human perception leads to implicit specifications of QoS requirements, especially in case of synchronization. For example, in case of a videoconference including a shared whiteboard, synchronization requirements are implicitly included. The pointer used in the whiteboard needs to move synchronous to the explanations delivered through the audio stream. These synchronisation issues [StNa96] need to be addressed explicitly at the application and network level.

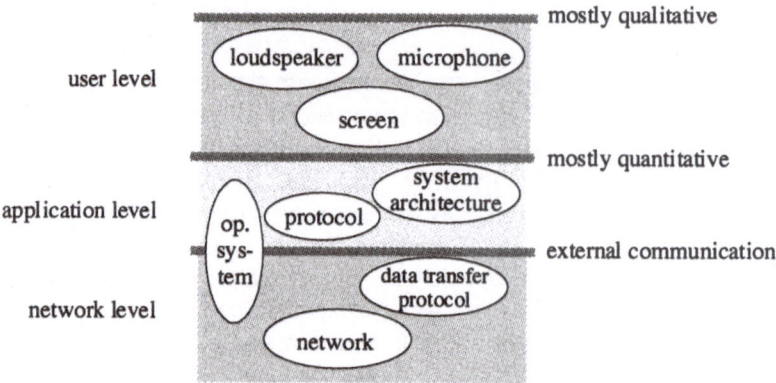

Figure 1 Model identifying QoS levels within a communication system

The user level comprises resources that are sources or sinks of data (camera, loudspeaker, screen, and the like). User level requirements need to be translated into application level requirements. Therefore, the communication system holds mapping tables which are used to transform the fuzzy and highly subjective and qualitative requirements to more precise and quantitative parameter values are needed at the application level interface.

The user level passes mostly quantitative parameters to the *application level*. Typical examples are the size of the application data unit, maximum data rate expected, and maximum accepted delay.

The *application level* deals with local parameters only. Resources at the application level are application-related protocols and protocol mechanisms, the operating system as well as the system architecture.

The operating system interacts with the *network level* due to its scheduling and memory management tasks. Interrupt handling is also an important issue that influences achievable QoS guarantees. The network level deals with parameters that are concerned with the quality of service of external communication. The network level interfaces to all functionalities that are mainly dependent on the network behaviour and that regulate the end-to-end data flow.

Between different QoS levels, the type of parameters changes. At the user level they are more or less solely qualitative parameters. At the application level, mostly quantitative parameters appear, mixed with few qualitative parameters. The network level deals with quantitative parameters only. These parameters are related to external communication. The network level QoS is directly comparable to end-to-end QoS often referred to in research papers.

Besides this three-level QoS model shown in figure 1, other approaches have been presented. For example, in [NaSt95] five levels are distinguished including, in addition to the three-level model, a system level and a device level. However, the distinction between user and device level is not needed if the model is viewed by a proper degree of abstraction. Both levels represent sources and sinks of data. Similar arguments hold for the distinction of application and system levels. Additionally, the same type of resource applies to different levels in the model, which does present a clean design. Using three levels seems to be a cleaner design than distinguishing many levels that only have marginal differences. Another approach of decomposing communication systems into different QoS levels is presented in [CaAH96]. The authors follow more or less directly the OSI reference model; they even add additional layers. A clear separation of tasks and resources is missing.

3. Emerging Internet Protocols

In order to support multimedia applications and scientific applications according to the service requirements presented in section 2, suited protocols are needed that implement these services. Current protocols basically do not consider QoS at all. Resource management and resource reservations being substantial to any kind of QoS guarantees are also missing.

In the context of the worldwide established Internet new protocols, such as RSVP (Resource Reservation Protocol), are under discussion for supporting the transfer of audio and video data (cf., figure 2) [Thom96]. Some protocols specifically focus on the support of multimedia applications and related QoS requirements. Popular examples can be seen in the network layer protocol RSVP and the application layer protocol RTP. Additionally, the network layer protocol IPv6 needs to be addressed which includes - in co-operation with RSVP - some features, namely flow labels, that can be used to support multimedia applications.

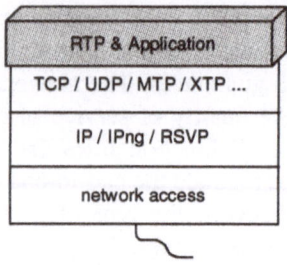

Figure 2 Current and emerging Internet protocols

Besides these protocols, various other protocol designs have been presented in current research papers, among them XTP [StDW92] and ST2+ [ST2+95]. Within this paper, we focus on Internet protcols due to their potential for widespread usage.

3.1 Future of IP

The new version of IP is called IPv6. The main motivation to develop a new network layer protocol originated in the nearly exhausted IP address space. Due to the use of longer IP addresses in IPv6 (128 bit instead of 32 bit), no further shortcomings in this direction are expected. In addition to changes in the addressing scheme, the structure of data units was adapted in order to allow for potentially more efficient header processing and for more service and protocol flexibility. The new structure of IPv6 data units (cf., figure 3) enables an easy integration of emerging IP options and, thus, provides a high potential for service flexibility. A basic IPv6 header is mandatory for every IP packet. It comprises the minimal information that is needed (e.g., source address, destination address, and hop limit). Additional functionality is carried in separate extension headers that are linked by pointers. Examples of such headers are fragmentation, specific routing, and security.

V	Pr	Flow-ID		1
length		pointer	Hop-Limit	2
source address				3..6
destination address				7...10

IPv6 header	Routing header	Fragment header	dara

Figure 3 Basic IPv6 header and IPv6 header list

IP packets that need to be served according to specific service requirements can be marked by a flow label. Flows can be established via a dedicated signalling protocol, such as RSVP. This way, support for continuous data flows can be achieved. However, the flow label can also be choosen randomly by the source in case that no dedicated service requirements exist.

Currently IPv6 is being introduced in an overlay network. This basically follows the positive experience that has been collected with the introduction of multicasting via the MBone overlay network. Some router manufacturers are working on the deployment of IPv6 implementations.

3.2 Resource Reservation: RSVP

The Resource Reservation Protocol RSVP is currently under discussion within the Internet [Brad95]. It is a signalling protocol which complements the data transfer protocol of the network layer (e.g., IPv4 or IPv6). The task of RSVP is to signal QoS requirements to the systems involved, i.e., routers and end systems. According to the QoS requirements, resources are reserved in the corresponding systems. These resource reservations form the basis to implement a guaranteed service requested by an individual application. Resources under consideration include network bandwidth, memory and processor capacity. Furthermore, specific support considering schedulers and the like is needed. Data transfer is implemented via IPv4, IPv6 or some other network layer protocol. RSVP data units are transported within IP packets. This is similar to the transportation of ICMP data units.

RSVP relies on existing or emerging routing protocols that have multicasting capabilities. It uses the routes provided by such protocols.

RSVP supports group communication with heterogeneous group members. It establishes reservations on a so-called multicast tree. Reservations are implemented in a receiver-oriented way with so-called RESV messages (cf., figure 4) in order to avoid temporary over reservations and, thus, unnecessary blocking of other requests. In contrast to that, ST2+ uses sender-oriented reservations. Two phases of reservation are needed: one phase from the sender to the receiver and the other phase vice-versa. In the first phase more resources may be reserved than actually needed. This can lead to the blocking of new reservation requests. Over reserved resources are freed within the second phase because then the end-to-end requirements (including the receivers service quality) are known.

RSVP uses so-called *flows* which can be seen as a sort of a soft connection. They are used in order to identify data flows with dedicated service requirements. Those flows are identified by flow labels that are also part of the IPv6 basic header (cf., figure 3). Thus, a flow reflects a sequence of data travelling from a source to a destination. No hard state information is associated with a flow, but so-called *soft states*. Flow establishment does not use handshake procedures but uses timer-based mechanisms. There are no acknowledgements required on changes in the soft states. However, the concept of soft states allows to implement soft guarantees only for multimedia applications. No hard guarantees can be given. Flows are located somewhere between traditional connection-oriented and connectionless services.

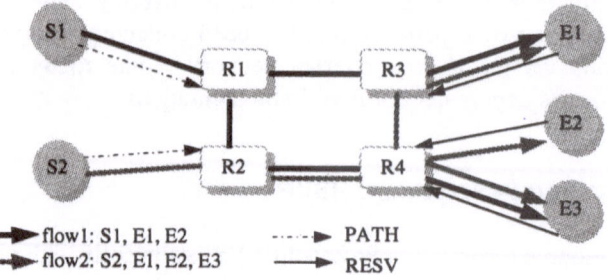

Figure 4 RSVP signalling

Reservations use flow descriptors which comprise a flow specification (flow-spec) and a filter specification (filterspec). The requested service quality is described by the flow spec. The filterspec defines the data units that are served according to that service quality. Furthermore, various reservation styles can be choosen (e.g., fixed filters, shared filters). A filter defines the relation between a reservation and a receiving entity. A reservation can be implemented for a dedicated receiver or for a group of receivers within a multicast communication. In case of shared filters, the reservations are shared among multiple flows. Audioconferences with usually only one speaker being active are a typical example for such reservations. In this case, usually only a single flow is actively using the reserved resources. Therefore, only reservations for that purpose are needed.

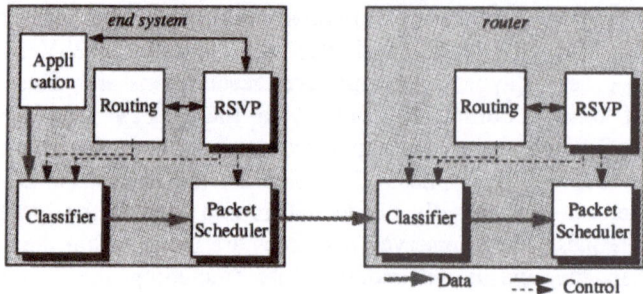

Figure 5 Implementation architecture of RSVP

The implementation architecture of RSVP is shown in figure 5. RSVP is located in end systems as well as in routers. An RSVP deamon resides at the user level. It is responsible for the signalling of QoS requirements. In the kernel space, a classifier and a packet scheduler are located. Data to be forwarded are first given to the classifier which determines the service quality needed as well as the subsequent route of the datagram. According to the service quality, the data are inserted into the queue of the output packet scheduler. The scheduler allocates the resources for transmission of the datagram. It is parametrized during the flow setup procedure.

3.3 Audio and Video Support: RTP

Another Internet protocol that was developed in order to support multimedia applications is RTP (Realtime Transport Protocol) [SCFJ95]. It is located at the application layer (not at the transport layer!) and dedicated to serve audio and video streams.

RTP provides sequence numbering of data and a time stamp needed for synchronization at the receiver (cf., figure 6). Group communication support is provided by RTP. Moreover, RTP can adapt to current network situations. Therefore, feedback information from the receivers is carried back to the sender. Based on this, the data rate may be adapted to the current load within the group. The feedback information is also delivered to all other receivers. Using such a mechanism runs into the same problem as end-to-end congestion control mechanisms. They need to address the possibly high latency until a feedback is collected.

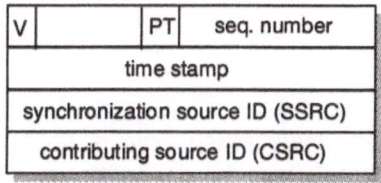

Figure 6 RTP header

RTP does not support any functionality that provides error control or error correction. Moreover, it does not support different quality of services and does not implement reservations. For these purposes, it relies on the underlying protocols of the network and transport layer. Usually, RTP is located on top of UDP. It is currently used in various applications, for example the network voice terminal NeVOT [Kuma95] and some of the MBone-Tools (e.g., vat and SD [Kuma95]).

Similarly to IP, RTP is composed of two components: RTP as a protocol to exchange real time data and RCMP (RTP Control Message Protocol) to exchange control information. Control information is related to quality of service and to the members of a group communication.

4. Group Communciation

Group communication is one of the paradigms that is inherent in almost all emerging networked applications. Both, multimedia applications as well as scientific applications can make use of this paradigm in order to accelerate themselves.

Within this section, focus is on the currently available group communication support within the Internet. Various multicasting protocols have been suggested for the transport layer, e.g., MTP (Multicast Transport Protocol) and its follow up MTP-II (BOGK94). The protocol XTP [StDW92] also includes revised multicast properties in its new version. However, none of them did yet qualify as *the* multicast transport protocol of the future. Therefore, they will not be discussed within this section.

It needs to be stated that besides the Internet developments, standardization activities are taking place within the ITU-T. In this context, the T.120 framework is devoted to the support of group communication. However, it is based on current transport protocols and, thus, on the use of point-to-point connections. Multicasting is emulated by the establishment of multiple point-to-point connections. A centralized Multipoint Control Unit (MCU) combines the various point-to-point links. Generally, the approach is not scalable to large and heterogeneous groups. Moreover, many tasks are delegated to the applications themselves.

4.1 IP multicasting

Within the last years, multicasting support inside the IP layer has been developed in the Internet, namely IP multicast. Currently a prototypical multicast network is exploited. It operates by using dedicated so-called multicast routers which form a multicast backbone (MBone) [Kuma95]. They use the so-called tunnel technique in order to interconnect multicast routers. This means, that routers encapsulate IPv6 packets into regular IPv4 packets that can be routed by the installed Internet routers. This way, a new technique can be introduced gradually. MBone forms the basis for experimental applications, such as video-conferencing and the like. Generally, MBone contributed highly to the success of emerging applications. This also holds for the amount of traffic introduced in the network (in certain parts around one third of the traffic). Generally, conferencing (and, thus, multicasting) is seen as the next major wave within the Internet.

Multicast IP supports the transmission of data units to a group of end systems. Specific IP multicast addresses are used. In addition to IP, the protocol IGMP (Internet Group Membership Protocol) [Thom96] which regulates the membership of a certain group is needed. Membership is regulated dynamically. However, the sender does not necessarily need to be member of the group. Moreover, the sender does not even know all the receivers at a certain point in time. A group can even consist of only one member. A single end system can concurrently be member of multiple groups. IGMP provides specific data units in order to join and leave a group. IGMP is located on top of the traditional IP protocol. With respect to the developments of IPng, functionalities of IGMP are migrated into the new Internet Control Message Protocol ICMPv6 [Thom96].

IP multicast provides an unreliable service for data delivery to a group of users. It does not include any error control or flow control mechanisms. This is subject of the transport protocol located on top of IP.

5. Implementation Techniques

With respect to system performance, implementation techniques of communication protocols play a major role. Processing requirements as well as data access frequencies (i.e., memory access and data copies) are rated as highly performance critical issues. Additionally, it is important to integrate the network attachment efficiently within the systems architecture in order to implement a high perfomance data flow (not only for large sizes of data units). As already stated, the requirement of low latency is common to multimedia and scientific applications. For example, in [AnCP95] they state that the latency of user-to-user communication on a workstation cluster should be less than 10 μs. In [MiBH95] even a latency of less than 1 μs is stated. One major portion of the overall latency, especially in case of small geographical distribution, is formed by the latency experienced within an end system. This includes processing overhead and overhead related to data movements. Especially the performance of scientific applications on workstation clusters may be dominated by this overhead. The following subsections describe different types of approaches that may improve the overall performance.

5.1 Network Adaptors

One approach to increase implementation efficiency can be seen in the use of special network adapters that largely avoid data movements. Figure 7 shows a typical scenario with multiple data copies (from the application buffer through the kernel buffer to the interface buffer). Moreover, the data are ususaly accessed during transport protocol operation in order to calculate the checksum. A good overview on different adapter designs is given in [Stee94].

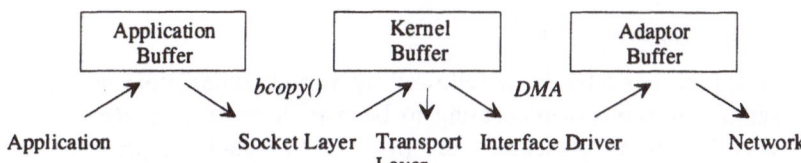

Figure 7 Data copies involved in data movements

Generally, memory accesses and data movements should be decreased to a minimum in order to allow for high performance implementations. In [EdMu95] and [MeMi94] an effort was made to implement a transport system with a minimum of data access operations. This is achieved by separating header information from user data. User data is kept untouched as long as possible. Some implementations have shown that it is possible to process network data on general-purpose workstations with line speed (up to 200-300 Mbit/s) if memory access is restricted to one copy operation.

An efficient buffer management is proposed in [DrPe93]. It is based on so-called *fbufs* and avoids cross-domain data copies. The U-Net proposed in

[EBBV95] also tries to minimize data movements. Moreover, kernel operation is removed from critical path operation, i.e., from user data transfers. This allows to streamline the buffer management according to the application requirements.

The network interface is capable of multiplexing and demultiplexing. Application data arriving at the interface are directly moved to its final destination at the user level. Such an approach is, for example, presented (cf., figure 8) in [MeZi95] and discussed in relation with ILP processing. Demultiplexing is based on the virtual channel identifier assigned to an application. This identifier is provided by the HP Jetstream which was used for the experiments. This approach targets protocol implementations to be part of the user space and to be more tied to the application itself. Received data are pre-processed at the driver level and copied directly into the user level memory. Protocol processing and de-/multiplexing is completely separated. Retransmission queues as they are needed at the sending side may be held at the user level or at board memory.

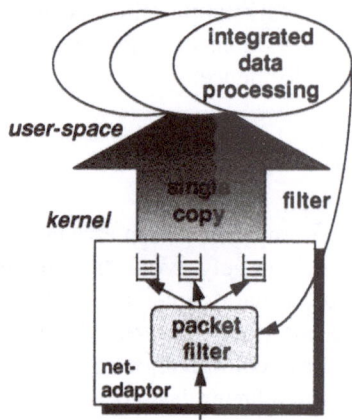

Figure 8 Minimizing data copies

Similar approaches have been followed by various other projects. An early project suggesting protocol processing to be part of the user space is the so-called packet filters presented in [MoRA87]. They implement some demultiplexing capability within the kernel and leave protocol processing to the user space. Generally, this leads to more flexibility. Flexibility in using different transport protocols and services is also among the goals of the approach presented in [TNML93] which favours protocol implementations at the user level. Furthermore, the project NOW (Networks of Workstations) [AnCP95] follows an approach that avoids kernel involvment at the critical path, i.e., for user data transfer.

An example of an efficient network access has been implemented at HP. With Afterburner they provide a network interface for the Jetstream LAN that is attached to the graphic bus of an HP workstation. Afterburner supports single copy implementations. The network adapter additionally provides hardware support for dedicated time-critical functions, such as checksumming. Another example of such a network adapter can be found in the Myrinet gigabit LAN [Bode95]. It provides a zero copy mode and supports checksumming on the adapter. The zero copy interface presented in [MiBH95] targets especially applications with very low latency requirements. They integrate the network interface into the memory hierarchy of the system. Thus, operating system involvement is avoided. The same can be achieved by the virtual memory mapped interface presented in [Blum95].

5.2 Integrated Layer Processing

Another more general approach for the preservation of memory access is known as Integrated Layer Processing (ILP) [ClTe90]. With ILP, all data processing is integrated into a single data processing loop. The goal is to integrate all processing that needs access to user data into a single loop and, thus, touch the data only once during their journey from the network to the application level and vice versa. Within ILP explicit data access for checksumming is avoided. Moreover, context switches are reduced. The ILP approach enables more efficient implementations, at least of simple communication systems. However, it also has some real drawbacks that limit the usage in general communication systems. Problems are created in case a non-linear operation of the protocol is required. Such an operation occurs if demultiplexing is implemented. It is clear, that the final destination must be selected before the ILP loop is entered. Examples for efficient implementations of demultiplexing have been presented in [CaJa93] and [ChVa95]. Reassembly creates problems similar to demultiplexing. The complete data unit must be available before entering the ILP loop. However, reassembly at lower layers is not possible, because state information from higher layers that caused the reassembly is needed. In order to allow reassembly, the function placement of traditional protocols needs to be adapted. Another typical transport layer function that causes problems during an ILP loop is retransmission. Packets to be retransmitted can only be stored at the entrance of a loop and not somewhere in between.

ILP is usually combined with a technique called Application Layer Framing (ALF). The ALF concept is based on the use of application level data units for protocol functions, such as error, rate and flow control. This avoids any problems concerned with segmentation and reassembly of data units within the ILP loop. Detailed experiments with ILP are presented in [BrDi96].

6. Summary

Various aspects of enhanced communication support for networked applications have been surveyed within this paper. Especially, services, protocols and implementation techniques are covered in some detail. These three aspects are very important in order to overcome current and emerging bottlenecks for both, multimedia and scientific applications. The overview given in this paper shows that a lot of progress has been made in enhancing communication support for such applications. However, there are still many open issues that need to be addressed in further detail. The approaches presented mainly reflect ongoing research work. Few of them are already implemented and widely used.

7. References

[Alle95] Anthony Alles; ATM Internetworking; Cisco Systems, May 1995

[Ande95] Th. Anderson, D. Culler, D Patterson; A Case for NOW (Networks of Workstations), IEEE Micro, February 1995

[ATTMF94] ATM Forum, ATM User Network Interface Specification, Version 3.1, July 1994

[ATMF95]ATM Forum, Traffic Management Specification Version 4.0, October 1995

[Bern96] P. Bernstein; Middleware: A Model for Distributed System Services; Communications of the ATM, Vol. 39, No. 2, Februar 1996

[Blum95] M. Blumrich, C. Dubnicki, E. Felten, K. Li, M. Mesarina; Virtual-Memory-Mapped Network Interface, IEEE Micro, Februar 95

[Bode95] N. Boden, D. Cohen, R. Felderman, A. Kulawik, C. Seitz, J. Seizovic, W. Su, Myrinet: A Gigabit-per-Second Local Area Network, IEEE Micro, February 1995

[BOGK94] C. Bormann, J. Ott, H. Gehrcke, T. Kerschat, N. Seifert; MTP-2: Towards Achieving the S:E:R:O: Properties for Multicast Transport, ICCCN, San Francisco, September 1994

[Brad95] R. Braden (Ed.); Resource ReSerVation Protocol (RSVP) - Version 1 Functional Specification; Internet Draft, July 1995

[BrCS94] R. Braden, D. Clark, S. Shenker; Integrated Services in the Internet Architecture: an Overview; RFC 1633, Juni 1994

[BrDi96] T. Braun, C. Diot; Performance Evaluation and Cache Analysis of an ILP Protocol Implementation; IEEE/ACM Transaction on Networking, Vol. 4, No. 3, June 1996

[CaAH96] A. Campbell, C. Aurrecoechea, L. Hauw; Architectural Perspectives on QoS Management in Distributed Multimedia Systems; Workshop on Quality of Service, Paris, France, March 1996

[CaJa93] S. McCanne, V. Jacobs; The BSD Packet Filter: A New Architecture for User-level Packet Capture; Winter USENIX conference, January, 1993

[ChVa95] G. Chandrammenon, G. Varghese; Trading Packet Headers for Packet Processing; ACM SIGCOMM, October, 1995

[ClTe90] D. Clark, D. Tennenhouse; Architectural Considerations for a New Generation of Protocols, ACM SIGCOMM, 1990

[DrPe93] P. Druschel, L. Peterson; Fbufs: A High Bandwidth Cross-Domain Transfer Facility; ACM SIGOPS, December 1993

[EBBV95] Th. Eicken, A. Basu, V. Buch, W. Vogels; U-NET: A User-Level Network Interface for Parallel and Distributed Computing, ACM SIGOPS, December 1995

[EdMu95] A. Edwards, S. Muir; A High-Performance Network Architecture for a PA-RISC Workstation; IEEE Journal on Selected Areas in Communications, Vol. 11, No. 2, February, 1995

[John96] H. Johnson; Fast Ethernet; Prentice Hall, 1996

[Kans96] T. Kanschik; QoS Monitor; Diploma thesis (in German), TU Braunschweig, 1996

[Kuma95] V. Kumar; MBone - Interactive Multimedia on the Internet; New Riders, 1995

[MeMi94] B. Metzler, I. Mioucheva; Design and Implementation of a flexible User Protocol Interface; Proceedings Hipparch, December, 1994

[MeZi95] B. Metzler, M. Zitterbart; Service Integration and ILP: Two contradictory Issues?, Hipparch Workshop, Sidney, Australia, December 1995

[MoRA87] J. Mogul, R. Rashid, M. Accetta; The Packet Filter: An Efficient Mechanism for User-level Network Code; Proceedings of the 11th Symposium on Operating System Principles, ACM SIGOPS, Austin, Texas, November 1987

[MiBH95]R. Minnich, D. Burns, F. Hady, The Memory-Integrated Network Interface, IEEE Micro, February 1995

[NaSt95] K. Nahrstedt, R. Steinmetz; Resource Management in Networked Multimedia Systems; IEEE Computer, May 1995

[NaSm95] K. Nahrstedt, J Smith; The QoS Broker; IEEE Multimedia, 1995

[ScZi95] C. Schmidt, M. Zitterbart; Towards Integrated QoS Management; LAN/MAN Workshop, Florida, 1995

[SCFJ95] H. Schulzrinne, S. Casner, R. Frederick, V. Jacoobson; RTP: A Transport Protocol for Realtime Applications; Internet Draft, March 1995

[ST2+95] Internet Stream Protocol Version 2 (St2), Protocol Specification, Version ST2+, Internet Draft, March 1995

[StDW92] W. Strayer, B. Dempsey, A. Weaver; The Xpress Transfer Protocol; Addison-Wesley, 1992

[Stee94] P. Steenkiste; A Systematic Approach to Host Interface Design for High Speed Networks; IEEE Computer, Vol. 27, No. 3, March 1994

[StNa96] R. Steinmetz, K. Nahrstedt; Multimedia: Computing, Communications and Applications; Prentice-Hall, 1996

[T120] ITU-T; Proposed Draft of T.120: Data Protocols for Multimedia Conferencing; März 1994

[Thom96] S. Thomas; IPng and the TCP/IP Protocols; John Wiley & Sons, 1996

[TNML93]Ch. Thekkath, T. Nguyen, E. Moy, E. Lazowska; Implementing Protocols at User Level; ACM SIGCOM 1993

[ZiST93] M. Zitterbart, B. Stiller, A.N. Tantawy; A Model for Flexible High-Performance Communication Subsystems; IEEE JSAC, 1993

[Zitt96] M. Zitterbart; User-to-User QoS: Management and Monitoring; IFIP Workshop Protocols for High Speed Networks, Sophia Antipolis, France, October 1996

List of Lectures

K. Abdali	*High Performance Computing Research under NSF Support*
G. Cooperman	*TOP-C: A Task-Oriented Parallel C Interface*
D. Du	*High Performance Computing over Switched-Based High Speed Networks*
S. Hariri	*High Performance Distributed Computing: Network, Architecture and Programming*
G. Havas	*Parallel Coset Enumeration*
A. Hoisie	*Very Large Scale Computations on the SP2: From network to application performance*
I. Janiszczak	*The Network of the Institute for Experimental Mathematics*
M. Mähler	*Switched Virtual Networking*
P. Martini	*ATM - The Networking Technology of the Future?*
A. Meyer	*Parallel Finite Element Simulation of Nonlinear Problems in Solid Mechanics and Fluid Dynamics*
B. Müller-Clostermann	*From Protocol Specification to Performance Modelling*
I. Niemegeers	*Signalling and Control for Multimedia Communication in Broadband ISDN*
I. Nikolaidis	*Discrete Event Distributed Simulation in Wide Area Networks*
G. Quintana-Orti	*Parallel Bidimensional Algorithms for Rank-Revealing QR Factorizations*
J. Rosenboom	*Computing with Modular Representations on Parallel and Distributed Computers*
G. Schneider	*ATM in Practice - Experiences in LAN and WAN Environments*
W. Schönauer	*Measurement, Proper Scaling and Application of Communication Parameters*
S. Srinidhi	*Geographically Dispersed computing over high-latency ATM links*
R. Staszewski	*Matrix Multiplication over Small Finite Fields on MIMD-Architectures*
X. Sun	*Parallel Dense Eigenvalue Solutions: Environments, Experiences and Expectations*
M. Weller	*Parallel Gaussian Elimination over Small Finite Fields*
M. Zitterbart	*Communication Support for Parallel and Distributed Applications*

List of registered participants

Dr. Kamal Abdali
National Science Foundation
1800 G Street
N. W. Washington DC 20550
USA
kabdali@ndf.gov

Prof. Gene Cooperman
College of Computer Science
Northeastern University
M/S 215 CN
Boston, MA 02115
USA
gene@ccs.neu.edu

Prof. David Hung-Chang Du
Department of Computer Science
University of Minnesota
Minneapolis, MN 55455
USA
du@cs.umn.edu

Prof. Dr. B. Fischer
Fakultät für Mathematik
Universität Bielefeld
Universitätsstraße
33615 Bielefeld

Dipl.-Phys. V. Gebhardt
Institut für Experimentelle Mathematik
Universität GH Essen
Ellernstr. 29
D-45326 Essen
gebhardt@exp-math.uni-essen.de

Prof. Dr. W.-D. Geyer
Mathematisches Institut der
Universität Erlangen-Nürnberg
Bismarckstr. 1 1/2
91054 Erlangen
geyer@mi.uni-erlangen.de

Prof. S. Hariri
Department of Electrical and Computer Engineering
Syracuse University
123 Link Hall
Syracuse, N. Y. 13244-1240, USA
hariri@cat.syr.edu

Dipl.-Ing. O. Harmjanz
Institut für Experimentelle Mathematik
Universität GH Essen
Ellernstr. 29
D-45326 Essen
harmjanz@exp-math.uni-essen.de

Prof. George Havas
Dept. of Computer Science
Universiy of Queensland
Queensland 4072
Australien
havas@cs.uq.oz.au

Dr. Adolfy Hoisie
Cornell University
623 Eng. & Theory Ctr Bldg
Ithaca, NY 14853-3801
USA
hoisie@tc.cornell.edu

Dr. I. Janiszczak
Institut für Experimentelle Mathematik
Universität GH Essen
Ellernstr. 29
D-45326 Essen
ingo@exp-math.uni-essen.de

Dr. M. Mähler
IBM Deutschland GmbH
European Network Center, Heidelberg
Vangerowstr. 18
69115 Heidelberg
maehler@ heidelbg.ibm.com

Prof. Dr. P. Martini
Fachbereich Mathematik und Informatik
Universität Paderborn
Warburger Str. 100
33098 Paderborn
martini@uni-paderborn.de

Prof. Dr. A. Meyer
Technische Universität Chemnitz-Zwickau
Fachbereich Mathematik
Reichenhainer Str. 41
09126 Chemnitz
a.meyer@mathematik.tu-chemnitz.de

Prof. Dr. G. Michler
Institut für Experimentelle Mathematik
Universität GH Essen
Ellernstr. 29
D-45326 Essen
archiv@exp-math.uni-essen.de

Prof. Dr. B. Müller-Clostermann
Fachbereich Mathematik und Informatik
Universität GH Essen
Universitätsstraße
45117 Essen
bmc@informatik.uni-essen.de

234

Prof. Dr. ir. I. G. M. M. Niemegeers
Tele-Informatics and Open Systems Group
Dept. of Computer Science
University of Twente
P. O. Box 217
NL-7500 AE Enschede
niemegee@ctit.utwente.nl

Prof. Dr. U. Stammbach
HG G 65.2 - Mathematik
Eidgenössische Hochschule
Zürich
CH- 8092 Zürich
stammb@math.ethz.ch

Dr. Ionais Nikolaidis
Assistant Professor
Computing Science Department
University of Alberta. Edmonton
Kanada
yannis@cs.ualberta.ca

Dr. R. Staszewski
Institut für Experimentelle Mathematik
Universität GH Essen
Ellernstr. 29
D-45326 Essen
reiner@exp-math.uni-essen.de

Prof. G. Quintana-Orti
Departamento de Informatica
Universidad Jaime I
Campus Penyeta Roja
E- 12071 Castellon
Spanien
gquintan@vents.uji.es

Prof. Xiaobai Sun
Duke University
Department of Computer Science
D107, Levine Science Research Center
Box 90120
USA-Durham, NC 27708-0129
xiaobai@cs.duke.edu

Dipl.-Ing. P. Roelse
Institut für Experimentelle Mathematik
Universität GH Essen
Ellernstr. 29
D-45326 Essen
roelse@exp-math.uni-essen.de

Prof. Dr.-Ing. H. Vinck
Institut für Experimentelle Mathematik
Universität GH Essen
Ellernstr. 29
D-45326 Essen
vinck@exp-math.uni-essen.de

Dr. J. Rosenboom
Institut für Experimentelle Mathematik
Universität GH Essen
Ellernstr. 29
D-45326 Essen
jens@exp-math.uni-essen.de

Dipl.-Math. C. Wagner
Institut für Experimentelle Mathematik
Universität GH Essen
Ellernstr. 29
D-45326 Essen
clemens@exp-math.uni-essen.de

Prof. Dr. G. Schneider
Universität (TH) Karlsruhe
Rechenzentrum . Zirkel 2
Postfach 6980
76128 Karlsruhe
schneider@rz.uni-karlsruhe.de

Dr. M. Weller
Institut für Experimentelle Mathematik
Universität GH Essen
Ellernstr. 29
D-45326 Essen
eowmob@exp-math.uni-essen.de

Prof. Dr. Schönauer
Universität (TH) Karlsruhe
Rechenzentrum . Zirkel 2
Postfach 6980
76128 Karlsruhe
schonauer@rz.uni-karlsruhe.de

Dipl.-Ing. A. van Wijngaarden
Institut für Experimentelle Mathematik
Universität GH Essen
Ellernstr. 29
D-45326 Essen
adraan@exp-math.uni-essen.de

Prof. S. Srinidhi
Sterling Software (US) Inc.
Federal System Group. MS 142-1
NASA Lewis Research Center
21000 Brookpark Road
USA-Cleveland. OH 44135
saragur@veda.lerc.nasa.gov

Frau Prof. M. Zitterbart
Institut f. Betriebssysteme u. Rechnerverbund
TU Braunschweig
Bueltenweg 74/75
38106 Braunschweig
zit@ibr.cs.tu-bs.de

Lecture Notes in Control and Information Sciences

Edited by M. Thoma

1993–1997 Published Titles:

Vol. 186: Sreenath, N.
Systems Representation of Global Climate
Change Models. Foundation for a Systems
Science Approach.
288 pp. 1993 [3-540-19824-5]

Vol. 187: Morecki, A.; Bianchi, G.;
Jaworeck, K. (Eds)
RoManSy 9: Proceedings of the Ninth
CISM-IFToMM Symposium on Theory and
Practice of Robots and Manipulators.
476 pp. 1993 [3-540-19834-2]

Vol. 188: Naidu, D. Subbaram
Aeroassisted Orbital Transfer: Guidance
and Control Strategies
192 pp. 1993 [3-540-19819-9]

Vol. 189: Ilchmann, A.
Non-Identifier-Based High-Gain Adaptive
Control
220 pp. 1993 [3-540-19845-8]

Vol. 190: Chatila, R.; Hirzinger, G. (Eds)
Experimental Robotics II: The 2nd
International Symposium, Toulouse,
France, June 25-27 1991
580 pp. 1993 [3-540-19851-2]

Vol. 191: Blondel, V.
Simultaneous Stabilization of Linear
Systems
212 pp. 1993 [3-540-19862-8]

Vol. 192: Smith, R.S.; Dahleh, M. (Eds)
The Modeling of Uncertainty in Control
Systems
412 pp. 1993 [3-540-19870-9]

Vol. 193: Zinober, A.S.I. (Ed.)
Variable Structure and Lyapunov Control
428 pp. 1993 [3-540-19869-5]

Vol. 194: Cao, Xi-Ren
Realization Probabilities: The Dynamics of
Queuing Systems
336 pp. 1993 [3-540-19872-5]

Vol. 195: Liu, D.; Michel, A.N.
Dynamical Systems with Saturation
Nonlinearities: Analysis and Design
212 pp. 1994 [3-540-19888-1]

Vol. 196: Battilotti, S.
Noninteracting Control with Stability for
Nonlinear Systems
196 pp. 1994 [3-540-19891-1]

Vol. 197: Henry, J.; Yvon, J.P. (Eds)
System Modelling and Optimization
975 pp approx. 1994 [3-540-19893-8]

Vol. 198: Winter, H.; Nüßer, H.-G. (Eds)
Advanced Technologies for Air Traffic Flow
Management
225 pp approx. 1994 [3-540-19895-4]

Vol. 199: Cohen, G.; Quadrat, J.-P. (Eds)
11th International Conference on
Analysis and Optimization of Systems –
Discrete Event Systems: Sophia-Antipolis,
June 15–16–17, 1994
648 pp. 1994 [3-540-19896-2]

Vol. 200: Yoshikawa, T.; Miyazaki, F. (Eds)
Experimental Robotics III: The 3rd
International Symposium, Kyoto, Japan,
October 28-30, 1993
624 pp. 1994 [3-540-19905-5]

Vol. 201: Kogan, J.
Robust Stability and Convexity
192 pp. 1994 [3-540-19919-5]

Vol. 202: Francis, B.A.; Tannenbaum, A.R.
(Eds)
Feedback Control, Nonlinear Systems,
and Complexity
288 pp. 1995 [3-540-19943-8]

Vol. 203: Popkov, Y.S.
Macrosystems Theory and its Applications:
Equilibrium Models
344 pp. 1995 [3-540-19955-1]

Vol. 204: Takahashi, S.; Takahara, Y.
Logical Approach to Systems Theory
192 pp. 1995 [3-540-19956-X]

Vol. 205: Kotta, U.
Inversion Method in the Discrete-time
Nonlinear Control Systems Synthesis
Problems
168 pp. 1995 [3-540-19966-7]

Vol. 206: Aganovic, Z.;.Gajic, Z.
Linear Optimal Control of Bilinear Systems
with Applications to Singular Perturbations
and Weak Coupling
133 pp. 1995 [3-540-19976-4]

Vol. 207: Gabasov, R.; Kirillova, F.M.;
Prischepova, S.V.
Optimal Feedback Control
224 pp. 1995 [3-540-19991-8]

Vol. 208: Khalil, H.K.; Chow, J.H.;
Ioannou, P.A. (Eds)
Proceedings of Workshop on Advances
inControl and its Applications
300 pp. 1995 [3-540-19993-4]

Vol. 209: Foias, C.; Özbay, H.;
Tannenbaum, A.
Robust Control of Infinite Dimensional
Systems: Frequency Domain Methods
230 pp. 1995 [3-540-19994-2]

Vol. 210: De Wilde, P.
Neural Network Models: An Analysis
164 pp. 1996 [3-540-19995-0]

Vol. 211: Gawronski, W.
Balanced Control of Flexible Structures
280 pp. 1996 [3-540-76017-2]

Vol. 212: Sanchez, A.
Formal Specification and Synthesis of
Procedural Controllers for Process Systems
248 pp. 1996 [3-540-76021-0]

Vol. 213: Patra, A.; Rao, G.P.
General Hybrid Orthogonal Functions and
their Applications in Systems and Control
144 pp. 1996 [3-540-76039-3]

Vol. 214: Yin, G.; Zhang, Q. (Eds)
Recent Advances in Control and Optimization
of Manufacturing Systems
240 pp. 1996 [3-540-76055-5]

Vol. 215: Bonivento, C.; Marro, G.;
Zanasi, R. (Eds)
Colloquium on Automatic Control
240 pp. 1996 [3-540-76060-1]

Vol. 216: Kulhavý, R.
Recursive Nonlinear Estimation: A Geometric
Approach
244 pp. 1996 [3-540-76063-6]

Vol. 217: Garofalo, F.; Glielmo, L. (Eds)
Robust Control via Variable Structure and
Lyapunov Techniques
336 pp. 1996 [3-540-76067-9]

Vol. 218: van der Schaft, A.
L_2 Gain and Passivity Techniques in Nonlinear
Control
176 pp. 1996 [3-540-76074-1]

Vol. 219: Berger, M.-O.; Deriche, R.;
Herlin, I.; Jaffré, J.; Morel, J.-M. (Eds)
ICAOS '96: 12th International Conference on
Analysis and Optimization of Systems -
Images, Wavelets and PDEs:
Paris, June 26-28 1996
378 pp. 1996 [3-540-76076-8]

Vol. 220: Brogliato, B.
Nonsmooth Impact Mechanics: Models,
Dynamics and Control
420 pp. 1996 [3-540-76079-2]

Vol. 221: Kelkar, A.; Joshi, S.
Control of Nonlinear Multibody Flexible Space
Structures
160 pp. 1996 [3-540-76093-8]

Vol. 222: Morse, A.S.
Control Using Logic-Based Switching
288 pp. 1997 [3-540-76097-0]

Vol. 223: Khatib, O.; Salisbury, J.K.
Experimental Robotics IV: The 4th International
Symposium, Stanford, California,
June 30 - July 2, 1995
596 pp. 1997 [3-540-76133-0]